RESEARCH ETHICS

D1041014

RESEARCH ETHICS

Cases and Materials

Edited by
Robin Levin Penslar

Indiana University Press

Bloomington and Indianapolis

© 1995 by Indiana University Press

All rights reserved

No part of this book may be reproduced or utilized in any form or by any means, electronic or mechanical, including photocopying and recording, or by any information storage and retrieval system, without permission in writing from the publisher. The Association of American University Presses' Resolution on Permissions constitutes the only exception to this prohibition.

The paper used in this publication meets the minimum requirements of American National Standard for Information Sciences—Permanence of Paper for Printed Library Materials, ANSI Z39.48-1984.

Library of Congress Cataloging-in-Publication Data

Research ethics : cases and materials / edited by Robin Levin Penslar.
p. cm.
Includes bibliographical references and index.
ISBN 0-253-34312-7.—ISBN 0-253-20906-4 (pbk.)
1. Research—Moral and ethical aspects. 2. Research—Moral and ethical aspects—Case studies. 3. Research—Methodology. 4. Research—Methodology—Case studies. I. Penslar, Robin Levin.
Q180.55.M67R46 1995
174'.90901—dc20 94-5971

1 2 3 4 5 00 99 98 97 96 95

For Derek, Joshua, and Talia

Contents

Contributors *ix*
Preface *xi*
Introduction *xiii*

Part I. Theory and Pedagogy

1 General Issues in Teaching Research Ethics 3

2 An Introduction to Ethical Theory 13

Part II. Cases in the Natural Sciences: Biology

3 The Professional Scientist 29

4 Scientific Misconduct: What Is It and How Is It Investigated? 41

5 Authorship and the Use of Scientific Data 51

6 Data Alteration in Scientific Research 56

7 The Ethics of Genetic Screening and Testing 62

8 Ethics and Eugenics 72

Part III. Cases in the Behavioral Sciences: Psychology

9 Ethical Issues in Animal Experimentation 87

10 Research Involving Human Subjects 99

11 Research Involving Human Subjects: The Administration of Alcohol 105

12 The Ethics of Deception in Research 112

13 Misconduct in Science 119

14 Science and Coercion 125

15 Behavior Control 137

Part IV. Cases in the Humanities: History

16 The Historian's Code of Ethics 147

17 The Use and Interpretation of Historical Documents 156

18 Oral Historians Meet the Media 160

19 Limiting Access to Scholarly Materials: The Case of the Dead Sea Scrolls 173

20 Faculty–Graduate Student Relations 182

21 Intellectual Property 187

Instructional Notes 193
Annotated Bibliography 239
Case Index 269
General Index 273

Contributors

Patricia Reuben Abreu
Graduate Assistant, The Poynter Center

Peter Cherbas
Associate Professor of Biology
Senior Fellow, Indiana Institute for Molecular and Cellular Biology

Peter Finn
Assistant Professor of Psychology

Tom Gould
Doctoral Candidate, Psychology

George Heise
Professor Emeritus of Psychology

Holly Irick
Postdoctoral Fellow, Biology

Herbert Kaplan
Professor of History

Douglas Lammer
Doctoral Candidate, Biology

Karen M. T. Muskavitch
Assistant Scientist, Biology
Fellow, Indiana Institute for Molecular and Cellular Biology

Lucinda Peach
Graduate Assistant, The Poynter Center

Robin Levin Penslar
Research Associate, The Poynter Center

M. Jeanne Peterson
Professor of History

Kenneth D. Pimple
Research Associate, The Poynter Center

David H. Smith
Professor of Religious Studies
Director, The Poynter Center

David Spaeder
Doctoral Candidate, History

William Timberlake
Professor of Psychology
Codirector, Center for the Integrative Study of Animal Behavior

Preface

THIS BOOK REPRESENTS the culmination of a project supported by the United States Department of Education's Fund for the Improvement of Postsecondary Education (FIPSE). It is fair to say that without FIPSE's support, for which I and the Poynter Center are most grateful, this book could never have been written. FIPSE's support enabled the Poynter Center to bring together faculty and graduate students from the departments of biology, history, and psychology at Indiana University to explore the ethical issues raised by the conduct of research in their respective fields and to develop teaching materials for both graduate and undergraduate students. The project was called "Catalyst" as a reflection of our hope that the project and its product would help stimulate discussion on research ethics at both Indiana University and institutions across the nation.

The cases reflect the combined expertise and sensibilities of all of the project's participants. While each case was initially drafted by a member of the particular field to which it relates, the effect on the final product of group discussions and editorial suggestions by various members precludes us from attributing individual authorship for specific cases. They are as much a reflection of the Catalyst group (as the participants came to be known) as they are of any particular member of the group.

I owe a debt of gratitude to the director of the Poynter Center, David Smith, for giving me the opportunity to explore the ethics of research with some of the University's premier scholars and teachers and their students. The participants in the project, most of whom had had no previous professional experience in working with these issues, were highly committed and gave of their time and energy far beyond any reasonable expectation. In addition to the Catalyst participants listed as contributors to this volume, many other Indiana University faculty, too numerous to mention here by name, provided invaluable comments and advice.

Various members of the Poynter Center staff provided their assistance during the course of the Catalyst project and the production of this casebook. My thanks to Lucinda Peach, Patricia Reuben Abreu, and Mary Mail. I owe special thanks to Kenneth Pimple for his help in preparing the final manuscript.

The office of Indiana University's Vice President for Research has been particularly supportive of the Center's efforts in the area of research ethics, of which the Catalyst project is but one example. My thanks to Vice President George Walker and Associate Dean Jeffrey Alberts for their continuing support, both financial and institutional.

RLP
Bloomington, Indiana

Introduction

THE CONDUCT OF research, whether it takes place in the university or in a for-profit venture, gives rise to a myriad of ethical questions. Ethical questions arise in all stages of research, from the design of the study, to the collection of data, to the reporting of findings in scholarly publications. Some issues, such as fraud, involve an intent to deceive—purposeful dishonesty—and are relatively straightforward. Others, however, are more subtle. What data should be included when reporting on one's research? When must data be shared? What is the relationship between student and mentor? What distinguishes contributions worthy of an authorship from those for which a footnoted acknowledgment is appropriate? When are deceptive research practices acceptable in research involving human subjects?

While these issues are by no means new (academics have grappled with them for generations), national attention has recently been drawn to the ethics of research by what seems like an alarming increase in the number of reported cases of research misconduct. Some of the cases have involved renowned academics—including Nobel laureates and important civic leaders—and research with important public health implications.

No one knows the incidence of inappropriate research activities, although the reported number of suspected instances of misconduct is certainly lower than the number that actually occurs. No solid studies have yet been published documenting the actual occurrence of research misconduct. We therefore have no idea what the incidence of questionable research practices is, or whether the rate has increased, decreased, or remained the same over the period of the perceived increase.[1] What we do know, however, is that the federal government is concerned enough about the perceived problem to seek ways to ensure that the research that it funds is conducted honestly. The Public Health Service now requires that every institution receiving PHS funds have in place procedures for handling allegations of research misconduct. In an effort to prevent the occurrence of research misconduct, various federal funding agencies (the National Institutes of Health and the National Science Foundation, for example) are working to encourage education in the responsible conduct of research.

At the same time as misconduct is a concern for the research establishment, so too are ethical concerns raised by the conduct of certain kinds of research. For example, the participation of human subjects in genetic research raises a host of ethical considerations. What are the risks to subjects of participation? Who has a right to access genetic information? How can subjects' identities be protected when pedigrees are published as part of a research report? Is it ethical for an investigator to gather information from a subject about that subject's family members if those family members have not consented to participation in the research? Further, what are the ethical implications of the human genome project (which seeks to map and sequence the human genome)? Many other kinds of research have the potential to affect a broad range of people and things. What ethical obligations does an investigator have to indigenous peoples who may be affected by his or her research? What are researchers' obligations toward the environment?

The need for education in the responsible conduct of research is not limited to graduate students or others who are embarking on research careers. We tend to think of research as something most relevant to the research community, to the academy. Research ethics, so the argument goes, is relevant to the education of future researchers. But while the importance of their relationship to the research enterprise and the obvious need to see to it that they are well trained in these matters cannot be stated too strongly, the importance of teaching research ethics to persons not headed toward research careers is equally great. We are all consumers of research; we must all understand not only the research enterprise, about which we presumably learn as part of our college education, but also its frailties.

The materials in this volume are designed to give faculty a vehicle for raising and discussing many of the ethically problematic aspects of conducting research. Since our goal was to provide materials for students on all postsecondary levels, the cases are directed toward university students on different levels, from undergraduates to postdoctoral fellows. The biology cases in chapters 5 and 6 and the materials on the use of historical documentation in chapter 17, for instance, were written specifically for use with undergraduates; the biology cases in chapter 3 and the history cases in chapters 20 and 21 were written specifically for graduate students and postdocs. Many of the materials, such as the psychology cases in chapters 11 and 13, are appropriate for students at any postsecondary level. When selecting cases for discussion, instructors will want to pay attention to the audience for whom the cases were written.

While ethicists will find the resources in this book useful, we intend by its publication to encourage faculty in other disciplines to discuss research ethics issues with their students, whether or not those faculty have had any formal training in ethics. Furthermore, the usefulness of the materials in this volume is not limited to the university environment. Scientists in industrial or other private laboratories or historians in corporate or government settings face many of the same issues as their counterparts in universities. These materials can be used to stimulate discussions in those laboratories or archives with the same goals and effects as in the university classroom.

As cases without facts would have little value, the materials are written with reference to the details of research in specific disciplines. We selected the disciplines of biology, psychology, and history as representative of the natural sciences, the behavioral sciences, and the humanities, respectively. But while they are presented in a subject-specific setting, many of the issues cut across disciplines. Plagiarism, for example, plagues all writers; animals are used in many of the biological and behavioral sciences; conflicts of interest can arise in any academic endeavor, as can problems of obtaining access to scholarly materials, be they chemical or documentary. We have found that the cases can be used "as written" in settings outside their natural context. For example, the biology cases, whose facts are drawn largely from molecular biology, have been used with great success with psychology and biology graduate students and postdoctoral fellows who work in animal behavior. The history cases on plagiarism have been used by faculty in English, law, and criminal justice. And the psychology cases on animals and on misconduct have been used in biology courses. In addition, faculty can and should adjust the facts to fit the particular course. The instructional notes provide ideas for how to work with specific cases in the classroom.

We hope and anticipate that this book will be used in a number of ways. The materials in this book can be used as part of a course in research ethics or as sources to stimulate discussion in a single class period focused on one or more ethical issues raised by a topic in the course. They can also be used as the basis for computer-based discussions among students outside of class or as the subject of a discussion section of a large lab class. They might also be used for writing assignments or provide the stimulus for students to write their own cases based either upon their own experiences or upon those they can envision from what they've learned of the research enterprise. Finally, they might serve as the subject of a graduate student roundtable, where, without the presence of faculty, graduate students are provided a forum for voicing their concerns and exploring their thoughts.

The book is divided into four parts. Part I focuses on theory and pedagogy. Chapter 1, "General Issues in Teaching Research Ethics," by Kenneth D. Pimple, introduces the reader to the two principal issues encountered by the new teacher of research ethics: why teach research ethics and what should be taught. Pimple then goes on to describe pedagogical strategies and techniques that underlie the teaching of practical ethics. In chapter 2, "An Introduction to Ethical Theory," Lucinda Peach provides an outline of the principal theoretical approaches to the analysis of ethical issues.

Parts II, III, and IV contain the case materials and questions for discussion. The cases are divided by discipline—biology, psychology, and history—and are grouped within each discipline by topic. Introductions to each group of cases provide the necessary background to the subject explored in that section. The cases cover a wide variety of topics, including authorship, confidentiality, conflicts of interest, fraud and misconduct, the obligations of mentors and teachers, and the reporting of data. In addition, the issues presented by the participation in research of human and animal subjects are explored. These issues include informed consent, coercion, risks and benefits, the participation of particular groups of subjects (such as students or alcohol-dependent individuals), and the care and use of animals.

Specific pedagogical suggestions for many of the cases and additional guidance for instuctors are provided in the Instructional Notes. These notes serve various functions, including providing additional issues that might be explored when teaching a particular case or set of cases. In some instances they also provide strategic pedagogical suggestions. They are to be distinguished from the Case Notes that appear at the end of several of the chapters. The Case Notes are adjuncts to the cases themselves, to be read as part of the chapter.

We have included an annotated bibliography on the ethics of research. This bibliography was developed to help identify resources to use as supplements to the cases, to assist readers who are developing courses in research ethics, and as an aid to further research in this area. Compiled by Lucinda Peach, it is divided into sections on biology, psychology, and history and subdivided by chapter and topic. An additional section provides references to sources of general interest.

Note

1. As this volume was going to press, Judith P. Swazey and colleagues (1993) published a long-awaited study on scientific misconduct. The study measures scientists' perceptions of the occurrence of research misconduct but does not document actual incidence rates. The authors questioned 2,000 faculty members and 2,000 graduate students in four scientific disciplines and found that a significant number of scientists had observed or had other direct evidence of colleagues engaged in inappropriate research practices.

Reference

Swazey, Judith P.; Anderson, Melissa S.; and Lewis, Karen Seashore. "Academic Research." *American Scientist* 81(6): 542 (November/December 1993).

PART I

Theory and Pedagogy

1 General Issues in Teaching Research Ethics

Kenneth D. Pimple

THERE SEEMS TO be a general feeling in the United States that unethical conduct is a large and growing problem, and there seems to be a growing recognition that this is not simply a personal problem but a social problem as well. Unethical conduct is a social problem not merely because it has a negative social impact but because social forces shape our personal ethical standards. There are at least two standards of right conduct in the United States: one is ethical or moral conduct,[1] and the other is success, whether measured in terms of wealth, popularity, power, or some other coveted social good. Ethical conduct and success are not necessarily at cross purposes, but they are often perceived to be, and we worry that the latter is likely to drive out the former.

Our teaching is designed to instill and reinforce the value of success; our students get grades, grants, degrees, and eventually jobs based upon their success as students. We also always teach something about ethics; some of us, unfortunately, teach that ethics does not matter much. Even a scrupulously ethical professor might implicitly teach that ethics does not really count if that professor never bothers explicitly to talk about ethics.

The common concern about ethics is frequently poorly articulated, but it seems clear that what we worry about most is practical ethics, that is, ethics as applied in practice, in everyday life, in government, journalism, medicine, education, in the professions generally. This volume is concerned with the teaching of ethics in one segment of professional conduct: scientific and other scholarly research.

My aim here is to provide some groundwork for the teacher of research ethics, a practical place from which to start. I point to three general issues that need to be addressed when teaching research ethics: ignorance, stress, and witnessing misconduct. It is common, I suspect, to address only ignorance, but the other two are important as well. I focus on the scientist or scholar[2] teaching research ethics, first discussing what we should teach, then offering a few thoughts on how to teach research ethics.

Scientists and scholars who are not trained as ethicists are often hesitant to take up the task. The reasons given vary widely. Some scholars claim that lack of training in ethics disqualifies them from the task of teaching ethics. Many scientists object that by the time students get to college, their character is formed and nothing that happens in any classroom can change them. Other teachers are pessimistic about human nature or about the younger generation. Although there are many nuances, the objections seem to boil down to two: I can't do it or it can't be done.

It is certainly difficult for someone with no training or background in ethics to teach research ethics. But it is easy to overemphasize both how little background a teacher has and how much is really needed. Many tools are available, including this volume. The real obstacle might be overly high expectations. If we think of the goal of teaching physics as making our students into physicists, then we might make the unfortunate analogy that the goal of teaching ethics is to make our students ethical. But we do not, in fact, have such lofty goals for a single course. In one course a teacher can teach *some* physics, and certainly in one course, or one course segment, even a nonethicist can teach *some* ethics.

The short answer to the second concern—that no one can do it—is that a growing body of empirical evidence shows that teaching ethics in college does make a difference in people's behavior (see Pascarella and Terenzini 1991).

If we accept that teaching ethics to young adults is both possible and desirable, the next question is how it should be done. Reticent scholars sometimes argue that it should be done somewhere other than in their classrooms or labs (at church or synagogue, for instance, or in a philosophy class). But we are talking about research ethics here, and teachers who want to shift the burden of teaching about the ethical practices and standards of their craft might well be ducking the responsibility, which seems in itself an ethically questionable move.

Research, like the professions, has its own peculiar demands, and in addition to a certain amount of crossover, specific disciplines face specific problems in research ethics. Philosophers or priests or rabbis might be most qualified to teach ethics broadly, but biologists and historians and engineers have the strongest responsibility to teach research ethics to the next generation of scholars and scientists.

What To Teach

One way to begin thinking about teaching research ethics is by thinking about ethical theories. But to the scientist or scholar who is well rooted in everyday life, distinctions between deontological (or duty-based) and utilitarian (or outcome-based) ethical theories might seem too abstract to be of much real use. Besides, the nonphilosopher faced with teaching research ethics might simply despair: I was not trained in ethical theory, so how can I possibly teach it?

Ethical theory has a place in teaching research ethics, but it is not necessarily the place to begin.[3] A different starting point is consideration of the kinds of ethical problems students are likely to face in their careers, first as students, then as scholars and scientists. As noted, these problems can be usefully divided into three categories: ignorance, stress, and witnessing misconduct.

The Problem of Ignorance

Some portion of unethical conduct stems from simple ignorance. To take an obvious example, some people simply do not understand how to cite sources properly, and they unknowingly plagiarize, perhaps by failing to use quotation marks properly or not using them at all. But not all examples are this obvious; the standards of science and scholarship are not self-evident, especially to novices, and it is the teacher's responsibility to pass on the standards of the discipline.

Other issues in research are less clear-cut than plagiarism. If I run an experiment one hundred times and the results fall along a neat line ninety-six times, can I just ignore the four outliers? Can I ignore five? Or six? Or ten? Or twenty? If I cannot simply ignore them, what am I to do with them? And how far out must a point lie to be an outlier?

Indeed, the issue of plagiarism itself is not as clear-cut as it once seemed to be. How much of a work must be quoted without citation to count as plagiarism? A paragraph? A sentence? A well-turned phrase? A colorful word? What about the sticky issue of plagiarizing ideas or concepts? How can we tell with certainty which ideas are truly in the common domain and which must be attributed? What is the role of intent? If I just forget to cite a source, is that plagiarism? Should negligence be included in the definition of misconduct?

One set of issues we must address, as tedious as it may seem, is the policies and standards of the field and of the university. I would not want to assume that action in accordance with policy is necessarily ethical action, and I would strongly resist the notion that action in accordance with policy is all there is to ethical action; but I would consider a scientist ill-trained who did not know the prevailing departmental, university, and federal policies on the use of human subjects, for example. If we teach nothing else when we teach research ethics (and I hope we can and will teach a great deal more), we must teach policies. In addition, though, we should talk about the spirit behind the policies, the principles that inform them and make them important.[4] We must also teach students where they can take their ethical concerns. Which dean deals with research misconduct?

If mentoring was ever sufficient training in the responsible conduct of research, it is not sufficient today (see Bird 1991). Labs are too large, deadlines are too tight, money is too scarce, and the issues are too complex and too subtle. Mentoring certainly plays a part; the example teachers set for their students shows clearly what they value and what they do not value. That should be a sobering thought, and it should make us examine exactly which values we model. It also points to the necessity for paying explicit attention to research ethics, for if we never talk about ethics, we may be teaching that ethical conduct is not important.

One (possibly unexpected) benefit of teaching research ethics is that it provides teachers an opportunity to clarify the policies, values, theories, and assumptions underlying their own practices. Any active research scholar or scientist deals with issues of attribution, authorship, ownership of research materials, and so forth, on a daily basis. As we all know, the best way to learn something is to teach it; and teachers of research ethics will learn to articulate what their own positions are and should be.

In a similar vein, talking about research ethics has a practical value for students. It lets them know what the issues are and where the teacher stands on them, what they need to grapple with, what can be negotiated, and what simply will not do. It may head off problems later, such as the filing of misconduct complaints by students over actions that are not unethical but are in fact the accepted standard of the field. Students also need to learn the proper standards in their work with other students, in attributing coauthorship, and in taking data with them when they leave a college or university.

The Problem of Stress

I have no empirical data, but I feel sure that the largest source of unethical conduct is simply stress: I must get a grant; I must meet a deadline; I must publish—now. In the face of the contingencies and exigencies of daily life and thrice-daily deadlines, just getting something done can seem more important than how it gets done.

We cannot remove all stress from students' lives, but we can teach them ways to deal with it. The most basic safeguard against stress-induced unethical conduct should be an ongoing part of all training: a thorough understanding of and commitment to the basic enterprise of science and scholarship—the pursuit of knowledge and truth.[5] What good is getting or keeping a grant if the data being produced are bad data? We can try to instill in students a sense of the futility of such actions. We already do instill it, I assume, through our actions, by modeling our own commitment to the pursuit of truth; but we can, and should, also take the time to talk about it, explicitly and more than once.

Ironically, a strong sense of the importance of the truth can itself give rise to stress and the potential for unethical conduct. Page limits make it impossible to "tell all" when we publish our findings, so we must pick and choose; and if we slice up our research into many articles in an attempt to tell the whole story, it might look like we are (unethically) trying to increase the number of our publications. What if I am "certain" that I know I have the right conclusion now and can save lives with it, even though all the data are not yet in?

Teachers can try to inculcate a long view. Cooking the books is tempting when we think of an experiment as the only one that counts, but is less tempting when we think in the long term. Falsified data provide only a foundation of sand, and the edifice built upon them will crumble one day, some way.

Related to this, we can teach patience, and the best way to do that is to talk to students about the importance of patience and to model patience for them, which means we must not always demand immediate results.

We can teach communication skills and encourage (or force) students to practice saying difficult things—like telling us teachers that we are demanding too much, that they just do not get it or just cannot seem to do it, or that our experiment was poorly designed and will not yield the kinds of answers we are asking for. Students can be taught that asking for help is better than cheating—but only by being treated decently when they do ask for help.

But the best way to deal with students' stress is to teach them to stop, take a breath, step back, and have a fresh look at the situation. Any one of us is capable of acting badly in the heat of the moment under circumstances which, if they were offered to us as a case study in the calm of the classroom, would not trip us up for a moment. We can teach students that when life gets stressful, they should try to take an objective look at their options. How would that course of action look if it were reported in the newspaper? Who will know, one hundred years from today, if I got the grant on the first try or the second?

We can also work toward implementing institutional and professional policies that relieve the stress that might lead to cheating, such as reducing the weight given to the number of publications in awarding tenure. In doing so, we hope that we will have a real impact, eventually; but if we let our students help us in that work, we can be sure that it will teach them a valuable lesson in responsibility.

Michael Josephson, founder and president of the Josephson Institute for the Advancement of Ethics, offers this bit of advice for people who think the only course of action open to them is an unethical one: "Take that solution you just said you have to do, and now make it impossible, for whatever reason. Do you now give up? Do you *die*? Is the world over? Or [do you] figure out something else? . . . [U]nless you have three alternatives to every major problem, you haven't thought hard enough. And as soon as you have three [solutions], you can find one of them that's ethical" (Josephson 1988).

It is precisely when life is stressful, of course, that we have a hard time sorting these things out, or even, more basically, finding the time to think about them at all. But we can teach students, in both word and deed, that this sorting out can and should be done and inculcate the habit of doing it in the relatively placid atmosphere of the classroom in the hope that it will follow them out into their own classrooms, laboratories, and businesses.

The Problem of Witnessing Misconduct

Even if all of our students always behave ethically, perhaps thanks to our tutelage, they will need to know what to do when they witness other people behaving unethically. This problem is clearly related to ignorance; it is unlikely that a person who does not know what is ethical can spot unethical behavior. But my focus here is on strategies: what do I do if I am a witness to unethical actions?

The first thing we can teach students about witnessing unethical behavior is that they must be sure of what they see. What we think is happening might not be what is actually happening; but it might be a graver error, in some cases, to assume the best of someone than it is to assume the worst.

It is also important to realize that the prudent and productive way to deal with suspected misconduct varies depending on many circumstances, not the least of which is the witness's standing vis-à-vis the suspected wrongdoer. Clearly, if I suspect person Alpha of wrongdoing, the actions I can take, the actions I must take, and the manner in which I take action vary considerably depending on whether Alpha is my student, my fellow postdoc, my peer in my department, my peer in another department at my university, my superior in my department, or a Nobel laureate. My personal relationship with Alpha might also make a difference; if we are amiable colleagues, I may be able to approach Alpha more straightforwardly than if we do not get along well or if we are direct competitors.

When we suspect students are acting unethically, our first duty is to instruct them, to teach them the error of their ways. If their unethical behavior is the result of ignorance, the solution might be as simple a matter as saying, "We do not do it that way; we do it this way." It is also important, however, to tell students why. When I was taught in freshman chemistry how to keep a lab notebook, I was told that I must keep it in pen, that I must never remove a page or add one, that I must take all my notes in that lab notebook, that I must never erase anything, and that if I crossed anything out I had to write a note explaining why I had done so. No one ever told me the reason for these standards, and I assumed it meant that in science we are not allowed to make mistakes. It would have been a simple thing for someone to tell me that these strictures are to ensure that all the data are kept together as accurately and completely as possible, and that I *was* allowed to make mistakes, but that I must record them because they contain important information; science is a process of learning from mistakes.

When students misbehave, we might have a duty to chastise them or punish them or report them, in addition to teaching them, but that varies from case to case; and even if we must punish students, our primary duty remains to teach them.

We should also prepare students for the possibility that they will witness irresponsible research behavior. It is difficult for some students to learn that the world is not an ideal place and that people who point out unethical conduct are sometimes punished rather than praised; it is our responsibility to help students understand this.

I would teach my students that caution is essential in dealing with any witnessed or suspected impropriety. I would teach my students to be doubly and triply sure of the facts, to walk extremely softly and wear self-righteousness lightly, to gather evidence quietly, to document abuses in writing, to try to find other witnesses and allies, and to have realistic expectations.[6] I would teach my students not to act at all until a firm foundation has been laid. Even so, the effects of the abuse must be considered; if life or health are at risk, there is a heavier burden on the witness to correct the situation, even at some personal cost.

Blowing the whistle is costly to everyone, and often, painful truth though it is, most costly to the whistleblower. I would advise my students that if they must face a suspected malefactor, it is best not to be self-righteous and confrontational. It is much better to say something like "Can you explain these anomalies to me?" than to say, "How long have you been forging data?"

Cases of suspected wrongdoing may unfold in many different ways. There might be a simple explanation and a simple solution; the suspected malefactor might admit to making a mistake but not to behaving unethically; he or she might deny that anything is wrong, technically or ethically. It might be possible to correct the error, but it might be necessary to go to higher authorities.

I would tell my students that if they witness unethical behavior on the job, somewhere along the line they should probably start looking for a new job, because it might be necessary for them to blow the whistle, and if they do, odds are they will need a new job.

I may be chided for being overly pragmatic and not sufficiently zealous in the pursuit of ethical conduct here, but making broad accusations of misconduct at too early a stage or in too rash a manner will not have a good effect on the accused and will have disastrous consequences for the accuser.

On the other hand, one could ask why, if blowing the whistle is so potentially costly, we should encourage students to take the risk. One simple reason is that in many circumstances it is a positive moral duty to right wrongs. Furthermore, we have a certain duty to uphold the integrity of our profession and our discipline. But on a pragmatic level, the misconduct policies of many institutions include a stipulation that faculty and students are required to report witnessed misconduct. It is important that students understand both the moral duty and the institutional policy.

As we teach research ethics, all of these possibilities should be considered and discussed overtly, and strategies can be mulled over in advance. In addition to strategies, of course, we must inform students of facts. Which dean deals with misconduct? What is the approved procedure for making a claim of misconduct? What are our policies? What are the regulations?

With luck, none of our students will need advice on whistleblowing; but if they need it, the effort in the classroom will be well spent.

How to Teach

In thinking about how to teach research ethics, it may be helpful to unpack the concept of "ethics" a bit. What does it mean to describe behavior as "ethical behavior"? James Rest has postulated that "four major kinds of psychological processes must have occurred in order for moral behavior to occur" (Rest et al. 1986: 3; see also Rest 1992). The four components are moral sensitivity, moral reasoning, moral commitment, and moral perseverance. In our teaching, we can address any one of these components, or any combination of them.

Moral sensitivity is simply the ability to recognize an issue or a problem as a moral problem. Many nonscientists would never guess at the complexity of the ethical issues surrounding research. We can help raise our students' moral sensitivity by explicitly identifying certain actions or ranges of actions as ethical and others as unethical and by discussing moral questions in research (such as the use of animals) even though the answers are not clear-cut.

Moral reasoning is the process of thinking about proper courses of action when faced with an ethical challenge. One of our goals must be to teach good reasoning skills, and reasoning about moral issues is not fundamentally different from reasoning about scientific issues. But in both cases, it helps to practice.

Moral commitment is what it takes to choose an ethically sound course of action over an unethical course of action. It is easy to imagine a person who can identify ethical issues with ease and can even determine what the most moral course of action would be but whose primary commitment is to some other, conflicting value, such as wealth or self-aggrandizement. We can help increase students' moral commitment by demonstrating to them that doing ethical research and doing successful research are, in almost all cases, synonymous.

Moral perseverance is having the ego strength and tenacity to follow through on one's decisions. It is similar to moral commitment but not identical. A person who lacks moral commitment is one who, in some circumstances, wants to act unethically because the unethical course of action is somehow more attractive than the ethical course; in contrast, a person who has strong moral commitment may still fail to act morally because of a lack of perseverance, an inability to face up to outside pressures and to actually do what that person genuinely wants to do and firmly believes should be done.[7]

In addition to showing students that responsible behavior in the conduct of research is important through our own actions and the way we treat them, one effective way to help develop these moral qualities is through the use of case studies in the classroom. One objection or fear about using case studies which seems to be common among new teachers of ethics is that when it comes to morality, there are no right answers. We live in a pluralistic society, and in some ways it seems that anything goes, that one person's opinion is just as good as another's. Teachers of ethics can understandably wonder, Who am I to judge my students' beliefs?

That is a complex concern. In many instances, it is true, there is no one right answer; but in every instance it is also true that there are many wrong answers, and in most instances a small subset of answers is better than all the other answers. Teachers may find occasionally that they and their students all agree on one single best course of action; but if they only find agreement that some courses of action are beyond the pale and others are clearly preferable, they will still have succeeded in teaching research ethics.

Cases can be used in many ways, some of which are discussed in the introduction to this volume. The aim can be to increase moral sensitivity by focusing on identifying the moral issues in a case or on improving moral reasoning by considering acceptable courses of action. These can also be combined in one exercise.

Research ethics can be taught in various formats ranging from full courses on the responsible conduct of research to one or more course units or discussion sections devoted to the topic. Appropriate courses for the latter include introductory courses, topics seminars, and research methods courses. The cases in this volume are suitable for use in any of these settings. Once teachers have gained some experience teaching cases, they may want to create cases tailored to their own needs and the needs of their students, perhaps drawing on the cases presented here for inspiration.

Teaching research ethics can be fun. Teachers who join with their students in thoughtful conversations about these difficult issues will find themselves learning about their students, and their students will learn about them. Teachers will discover perspectives they had not considered (as will students) and, in all likelihood, will learn a thing or two about the ethics of research even as they teach it.

Every good scholar and every good scientist who teaches research methods can and should teach research ethics. You may fear that you do not know how; I believe that that fear says more about your understanding of the complexity of the issues than your inability to teach them. If you are a responsible scientist or scholar, you must know how to conduct research responsibly, and all you need to teach it is some reflection on what you do anyway (making your practice explicit to yourself so you can make it explicit to your students), a little self-confidence, and a few tools, such as the cases and bibliography in this volume.

People make mistakes, and not all mistakes are evidence of unethical behavior. Crisp distinctions between unethical behavior and simple error are not always possible in theory, and even less so in life. How should we respond when we see students making mistakes? The same way we should respond when we see them engaging in unethical behavior—we should educate them, train them, and help them practice doing it right.

Notes

1. For the purposes of this essay, the words *moral*, *ethical*, and *responsible* should be considered roughly synonymous.

2. I do not mean to imply that scientists are not scholars; I mean to make it explicit that I include researchers in the humanities in this essay. I sometimes use *scientist* and sometimes *scholar* to cover scholars both in the sciences and the humanities. I also use the term *scientist* broadly to include anyone involved in the empirical and quantifiable study of the world—including, for example, engineers.

3. Lucinda Peach's "Introduction to Ethical Theory" (chap. 2 in this volume) provides a starting point for nonethicists.

4. Of course there are stupid policies, and there are counterproductive policies; but even those should be followed on pragmatic grounds, even as we work to overturn them. The existence of a few backward policies should not prevent us from teaching the wisdom of the wise policies.

5. The "truth" is, of course, a highly contestable category; but whatever the truth is, it is *not* the product of intentional duplicity, cooked books, or conduct we would generally call unethical.

6. Some of these ideas are drawn from "Ten Steps for Effective Whistleblowing," compiled by Donald R. Soeken, director of Integrity International, 6215 Greenbelt Road, College Park, MD 20740.

7. Richard L. Morrill divides the moral universe in a slightly different way, distinguishing between values analysis, values consciousness, values criticism, values pedagogy, and values development. See Morrill 1981, esp. chap. 4.

References

Bird, Stephanie J. "Education in Research Ethics: Report to the Committee on Academic Responsibility." Massachusetts Institute of Technology, August 1991.

Josephson, Michael S. *Michael Josephson.* Videorecording. Alexandria, VA: PBS Video, 1988.

Morrill, Richard L. *Teaching Values in College.* San Francisco: Jossey-Bass, 1981.

Pascarella, E. T., and Terenzini, P. T. *How College Affects Students.* San Francisco: Jossey-Bass, 1991.

Rest, James, with Bebeau, Muriel, and Volker, Joseph. *Moral Development: Advances in Research and Theory.* New York: Praeger, 1986.

Rest, James R. "A Psychologist Looks at the Teaching of Ethics." *Hastings Center Report* 22(1): 29–36 (January/February 1992).

2 An Introduction to Ethical Theory

Lucinda Peach

You discover that a colleague at State University, Dr. X, has published an article containing erroneous information. You are uncertain whether X intentionally or negligently included the erroneous information. In either event, the misinformation is significant in your field of research and is likely to send other researchers down unproductive paths. You are aware of no other problems with Dr. X's research, and you know that he is up for tenure next year. What do you do?

ETHICS (derived from the Greek *ethos*, meaning character, custom, or usage), or morality (from the Latin synonym meaning manner, custom, or habit), is the philosophical study of normative behavior, the "shoulds" and "oughts," the "rights" and "wrongs," of our conduct. Research ethics is a kind of applied or practical ethics, meaning that it attempts to resolve not merely general issues but also specific problems that arise in the conduct of research. Its goal is to determine the moral acceptability or appropriateness of specific conduct and to establish the actions that moral agents ought to take in particular situations. Research ethics is therefore not merely theoretical. It aims to establish practical moral norms and standards for the conduct of research.

The line between doing good research and doing ethical research is not always evident. Certain ethical considerations, such as honesty in gathering and reporting data, are integral to the practice of research. They are part of the standards and norms that comprise the scientific method. Other ethical considerations, such as avoiding harm to human subjects or exercising social responsibility for the consequences of one's research, are important, but they are outside the scope of what is required by established standards for determining good research. A few generalizations can be made, however, to distinguish ethical from nonethical considerations in research.

One way of drawing the distinction is to say that the conduct of research concerns the integrity and intellectual soundness of data, whereas research ethics relates to the means by which scientific data are obtained and the social consequences of their discovery and analysis. A second way of making the distinction is to say that ethical issues in research generally concern attention to the welfare of others, whereas standards governing the correct procedures of research do not necessarily do so. The relevant others may include particular individuals and animals as well as society as a whole (see Beauchamp and

Childress 1989: 14–17). A rule prohibiting sexual relationships between researchers and their subjects, for example, may or may not be necessary to ensure good research, but it is certainly an important component of ethical research.

Much of what has been formulated under the rubric of research ethics has focused on the conduct of individuals. Typical issues involve questions of research fraud and misconduct, mistreatment of laboratory animals and human subjects, accuracy and honesty in conducting and reporting research results, plagiarism, violations of intellectual property rights, and conflicts of interest between duties of researchers to their universities or other employers and rights to profitable use of the fruits of their research. But there is also a public policy dimension to research ethics that involves principles applicable to the professional conduct of science as a whole. Public policy issues involved in research ethics include determinations about whether a research project violates basic moral principles or is predictably harmful to identifiable populations, questions about whether the research is so controversial that it should be forbidden, what the scope of scientific responsibility to the public should be, how universities should regulate conflicts of interest between themselves and private commercial interests, and how they should evaluate research misconduct cases on their campuses.

A number of different sources for ethical norms or standards are applicable to research ethics. An important source governing much research is professional codes, statements of moral norms expressed by members of a profession. Examples are the codes of professional responsibility of the American Psychological Association and the American Historical Association. However, because professional codes are developed by members themselves for self-governance, do not reflect the views of clients or nonprofessionals, and are not required of the profession by an outside body or agency, they may not be an adequate statement of the rules and principles applicable to the moral conduct of that profession.

Governmental regulations are another source of moral norms. In addition to their binding force as law, such regulations often reflect moral assessments about the appropriateness of certain conduct. The federal regulations governing the involvement of human subjects in research provide a good example. The development of the human subjects regulations was explicitly based on the application of such moral principles as respect for persons, beneficence, and justice. In addition, particular cultural customs, norms, and standards, including those derived from religious and philosophical sources, may provide guidance in resolving ethical issues arising in the context of research.

Ethical Theories

Ethical theories can also be used to evaluate the morality of research activities. No single theory or approach to ethics is ideally or completely suited to resolving all issues that arise in the course of research; nonetheless, certain approaches may be more useful or effective than others, depending on the type of issue or problem involved.

In general, the two most common contemporary approaches to Western ethics can be described as *deontological* and *consequentialist*. In consequentialist theories, the rightness or wrongness of an action is determined with regard to the action's results. Perhaps the best-known consequentialist theory is utilitarianism, formulated by Jeremy Bentham and James and John Stuart Mill in the mid-nineteenth century, in which actions that lead to the greatest good for the greatest number of people are considered to be the best actions. By contrast, deontological theories focus primarily on some value such as human autonomy and only secondarily on outcomes. The best-known deontological theorist is Immanuel Kant. To a utilitarian, the best course of action might sometimes be to tell a lie (we can all imagine circumstances in which lying is the easiest way to make everyone happy), but a deontologist might say that telling a lie violates another person's autonomy so severely that lying is never justified.

These two types of ethical theory have dominated in the Western world since the nineteenth century. But alternative approaches to ethics are now being recognized or recovered, including those that focus on the particular facts of the situation involving a moral dilemma (*situation ethics*), on analogies between clear-cut and ambiguous cases (*casuistry*), or on the character of the actor rather than the actions themselves (*virtue ethics*). The remainder of this chapter will describe each of these approaches to ethics and how they have been or might be used in the context of academic research. The discussion of each approach starts with the application of the theory to the hypothetical problem described at the beginning of the chapter. All of the approaches involve paying attention to some or all of six factors: (1) facts, (2) interpretations of the facts, (3) consequences, (4) obligations, (5) rights, and (6) virtues.

Consequentialist Ethics

Under a consequentialist approach to ethics, you, as the moral agent,[1] ask: In general, what are the possible harms that may result from Dr. X's erroneous publication, and what are the harms that may result from the various ways I can respond, knowing that the publication is erroneous? Which response will result in more good than harm? More specifically, what harm might X's erroneous research results cause if not corrected? How does this harm weigh in balancing the harm that may occur to X personally if I bring the discrepancy in his data to the attention of others? What are the possible benefits of talking to X first, before taking further steps, and how do those benefits weigh against the possible harms to State University and the larger scientific community? What if X is guilty of deliberately publishing erroneous results, and my tip-off allows him to attempt to cover up his tracks? How will each potential course of action affect X if he is innocent (or at least did not deliberately publish erroneous data)?

Sometimes referred to as utilitarian or teleological (the latter from the Greek *telos*, "goal"), consequentialist theories are focused on the result or outcome of actions. In contrast to deontological ethics, the proper course of action is determined in accordance with its likely consequences or outcomes rather than its inherent rightness or wrongness. Consequentialist conclusions that are especially based on an impartial consideration of the interests or welfare of others are called utilitarian theories. The principle of utility states that we should strive to create the greatest possible balance of good over evil in the world. Utilitarianism is concerned with promoting human values, such as happiness (or other goods), by maximizing benefits and minimizing harms, as reflected in the maxims "the ends justify the means" and "the greatest good for the greatest number."

These ends or outcomes may refer to values such as happiness, health, knowledge, self-realization, perfection, or general welfare—outcomes which are intrinsically valuable, that is, desirable in and of themselves. These outcomes are nonmoral because one can have them as a matter of luck or fortune, without any moral effort or desert. But they are good things to have, however one comes to have them. The different utilitarian approaches are distinguished from one another depending on whether they measure this "good" in terms of promoting pleasure and avoiding pain (*hedonic* or *hedonistic utilitarianism*), satisfying preferences (*preference utilitarianism*), satisfying people's long-term interests (*welfare utilitarianism*), or as best satisfying the total range of intrinsic values, including friendship, knowledge, and health, in addition to happiness (*pluralist utilitarianism*).

Consequentialism provides a distinctive rationale for what might otherwise be labeled unethical conduct, by explaining it instead as conduct that fails to lead to the best consequences or outcomes under the circumstances. The order of priorities of utilitarian theories may thus be characterized as the "good" being prior to the "right." In addition, whereas deontology frequently focuses on the duties of one person in one situation, consequentialism generally concerns all those foreseeably affected by an action.

Finer distinctions among utilitarian approaches also exist. Utilitarian approaches are sometimes distinguished on the basis of whether they evaluate the consequences of particular acts (*act utilitarianism*) or the consequences of following general rules governing acts, such as "don't steal" (*rule utilitarianism*).

The act form of utilitarianism proceeds by attempting to assess which of the available courses of action available to the moral agent is likely to produce the greatest good in the particular situation at hand. Act utilitarianism is focused on a specific actor addressing a concrete and specific situation, not on what would generally be the best course of action for someone in that kind of situation. Moral principles, such as "telling the truth is generally for the greatest good," are relevant only for providing general guidance. In contrast to deontological approaches, they have no binding force or effect on the moral evaluation. Act utilitarianism has been criticized as cumbersome and inefficient, since it requires every moral actor to balance goods and harms in each and every moral situation encountered, despite similarities among cases.

Rule utilitarianism emphasizes the centrality of rules to morality and states that in general, one should follow those moral rules which result in providing the greatest utility for everyone. In contrast to act utilitarianism, rule utilitarianism is most concerned with finding and applying the rules, rather than the acts, that lead to the greatest utility. The rule utilitarian asks about the consequences of following a rule in a type of situation rather than being concerned with a particular act in a specific situation. Under rule utilitarianism, the moral agent will follow a rule that may not lead to the greatest utility in the particular situation, although it does in general. As a trade-off for this negative consequence, rule utilitarianism provides a solution to the inefficiency of act utilitarianism by providing some regular guidelines. These guidelines free moral agents from having to calculate utility in every situation, regardless of how many times they, or others, have encountered the same or similar circumstances. An act is judged right or wrong based on its conformity to a rule rather than on the act's beneficial consequences in a particular situation. One criticism that has been made of rule utilitarianism is that it may result in sacrificing justice in particular situations in order to promote the greatest general utility.

Moral judgment using either of these utilitarian approaches to ethics depends on a clear and comprehensive understanding of the facts involved in the particular moral dilemma or controversy. This is particularly important in the context of research ethics, as illustrated by the hypothetical case of Dr. X, where moral culpability may depend on whether the research-as-reported corresponds accurately with the original data.

Utilitarian theories have been criticized generally for failing adequately to ensure justice in the course of maximizing good over evil. Deontological theories are often counterposed as a way of successfully dealing with this problem.

Deontological Ethics

According to this second type of ethical approach, you, the moral agent, want to know: What is Dr. X's duty in publishing research, and what is my duty now that I have discovered errors in his article? What is the right thing to do in this case? Assuming X's duty is to report his findings accurately, a key question is whether X's paper deliberately or negligently included erroneous data. If X was negligent, a question remains about the best way to reverse the effects of the misreported information. If X deliberately misreported the results of his research, the moral agent may have a duty to report him or otherwise see to it that a process of investigation is undertaken.

Deontological ethics (from the Greek *deon*, meaning "duty") holds that some acts are intrinsically right or wrong, regardless of the consequences. Deontological approaches are rule-based (as are some forms of consequentialism); that is, particular judgments about how to resolve an ethical dilemma are supported or justified by reference to rules. The rules that are the basis of deontological ethics are justified by principles. The distinction between rules and principles is basically a matter of degree: rules are usually more specific and limited in scope than principles, which tend to be more general, broad, and abstract. For example, "don't deliberately distort your data by eliminating outliers" is a moral or ethical rule, whereas "truthfulness" is an ethical principle. Basic principles relevant to research ethics include truthfulness, fairness, justice, respect for persons and their intellectual property, and integrity.

In contrast to the consequentialists, deontologists (at least some of them) view moral rules as binding regardless of the consequences. Certain acts are considered to be wrong in themselves and thus morally unacceptable, regardless of the importance or moral commendability of the ultimate ends that they may be intended to achieve or that may actually result. Deontologists thus will do what is "right," determined in accordance with the constraints of rules, laws, prohibitions, prescriptions, and norms, regardless of the consequences. Another way of making this distinction is to say that deontology places a primary value on doing what is "right," whereas consequentialism accords priority (or place) to doing what will achieve the greatest "good."

A researcher taking a deontological approach to resolving a dilemma about reporting data will refuse to lie about research results even though slightly misrepresenting the data might more fairly characterize the overall trend of the findings. At its most extreme, deontology prohibits violating established standards of what is "right" according to the rules, even if doing so would result in the "good" of saving human lives. Deontologists contend that there are other values that take precedence over maximizing the balance of good over evil in the world.

Deontological constraints are typically formulated as prohibitions. They are also bounded, so that what lies outside the constraint is not prohibited. If the relevant standard for plagiarism is intentional quotation of another's work without attribution, then negligent failure to cite the quoted work does not constitute plagiarism. A critical distinction for the deontologist is between the intentional and the accidental or inadvertent. Since lying involves a direct and deliberate deception rather than an indirect and not necessarily deliberate one, a deontological approach would prohibit deception of the former sort but not of the latter.

Because deontologists refuse to judge the ethical permissibility of actions simply on the basis of their consequences, they may have a more difficult time than consequentialists in offering full explanations of certain acts as right or wrong. After all, rules may be bad, immoral, wrong, unjust, or impoverishing to human life. Many deontologists appeal to a tradition, particularly that of the Western biblical tradition, as a source for correct moral norms. Others rely on the Kantian Enlightenment principle of universal respect for persons or on an appeal to moral intuitions.

Rule-based approaches have been criticized as assuming that there is one "right" answer for every moral dilemma. In the absence of an established set of standards or a framework establishing which rules and principles take priority over others, it may be difficult to determine how to proceed. Rule-based approaches to ethics may require that the moral agent determine which of several rules or principles is most suitable or relevant to a particular situation.

Deontology has also been criticized for its legalistic and narrow application of norms without reference to their real-life consequences. Sometimes it may lead us to focus on the wrong thing. If failure to tell the whole truth has the same consequences as deliberately lying and is done with the same motivations, why should it not be judged to be equally unethical? Similarly, why should respect for persons be a more important basis for ethics than furthering the welfare of persons? Both consequentialism and deontology begin from certain premises about fundamental values, and justifications for those premises will take us far afield. Suffice it to say that these sorts of problems have led critics to label deontology as overly formalistic and legalistic—as encouraging only narrow and minimal adherence to ethical constraints rather than encouraging compliance with the spirit and not merely the letter of ethical standards.

Another objection to deontological approaches is that duties sometimes conflict: my duty to tell the truth and my duty to avoid harming others conflict with each other whenever telling the truth causes harm. One way deontologists have surmounted this argument is to distinguish between an actual duty and a prima facie duty. The former is what we actually ought to do in a given situation; the latter is what we ought to do if there are no intervening circumstances which may create problems for the actor in complying with the prima facie duty. Whereas the former cannot be stated in the form of exceptionless rules, the latter can be. When there are conflicts among prima facie duties found in different general rules, say between justice and beneficence, an agent's actual duty is determined by weighing the balance of all the competing prima facie duties. Although a prima facie duty may be overridden, it does not disappear but leaves "moral traces" that should lead the moral agent to regret or have remorse for not being able to fulfill the duty (see Beauchamp and Childress 1989: 52–53).

Rights are an important consideration in some deontological theories. They also are an important aspect of many issues in research ethics, including the rights of human research subjects to be treated with respect, to be able to grant or refuse consent to participation in research, to be guaranteed confidentiality, and to receive a debriefing following participation in experiments. Rights are relevant to determinations of authorship, ownership of intellectual property, and conflicts of interest. Some ethicists extend consideration of rights to animals involved in research.

The philosophical recognition of rights goes back to classical Greek notions of "obligations" founded on natural law and of claims founded on "natural rights" emerging from this natural law, which many Greeks, particularly the Stoic philosophers, viewed as transcending man-made laws. Whereas the eighteenth-century understanding of rights was generally limited to negative rights, which established limits on governmental interference with subjects, contemporary principles of rights include positive elements, such as rights to health, education, and welfare.

While legal and moral rights may be similar, they are not necessarily synonymous. A research colleague might have a moral right to be recognized as a contributor to a research publication but not a legal right to be listed as a coauthor. In addition, the notion of rights and duties may make sense only in relation to human conduct and not in relation to natural events in the world. It is not clear, for example, that a child has a right not to get leukemia.

Rights alone cannot form an adequate framework for ethics. Rather, they form one important element, usually correlating to corresponding duties of some other person. Rights therefore presuppose social responsibility, since the assertion of rights by one person necessarily involves recognition of the rights of others.

Not all deontological theories are rule-based. Act deontologists (or situationists) take particular judgments, rather than general rules, to be basic in morality. They hold that morality can only be assessed within the context of particular factual situations and not in accordance with general rules. If they recognize the existence of general rules at all, it is in terms of inductive generalizations from particular cases. Taken to its extreme, act deontology maintains that every moral decision must be determined in and by itself, without reference either to general rules or to calculations of utility.

Situational approaches that measure particular acts in accordance with one overarching general principle, such as love, may be viewed as subsets or types of act deontology. They share a central concern with the facts of particular, concrete moral dilemmas: what happened, when, involving whom, for what purpose, and with what consequences. Situational approaches may use rules, but only as guidelines, or rules of thumb. According to the situationist, there is only one very general standard that can always be applied to moral problems without reference to the particular facts involved. Guidelines can be dispensed with in particular situations, since they are not hard-and-fast prescriptions that are universally applicable to similar situations.

Act deontology (or *situationism*) has been criticized as failing to provide any standards for determining what is right or wrong in particular cases, since it holds that particular judgments take precedence over any general rules. On this view, although act deontology is commendable for recognizing the uniqueness of every factual situation, it is overly narrow in failing to recognize that many moral situations do have relevant commonalities that can be treated alike using general moral rules. Furthermore, it has been claimed, act deontology fails to recognize the principle of universalizability which is applicable to moral rules. Most simply, this principle holds that making a moral judgment in one set of circumstances commits the moral agent to making the same moral judgment in all other circumstances that are morally similar. Of course, there is often dispute about what precisely is intended by the term *morally similar*. In addition, critics contend that we cannot live without at least some rules to govern our conduct. Finally, act deontology has been criticized for failing to provide reasons for moral judgments. Moral and value judgments, it is claimed, imply reasons, and these reasons apply in all relevant cases, not only to one particular situation.

Casuistical Ethics

In accordance with the casuistical approach to ethics, you, the moral agent, ask: What are the paradigmatic cases that provide the outer boundaries for assessing Dr. X's erroneous publication? Is X's case more like a case of intentional wrongdoing, where the researcher has deliberately published false information for personal gain? Or is it more like a case of negligent oversight, where the researcher has failed to recheck the numbers in his or her data set carefully enough before submitting them for publication? If X's situation is closer to the former case, then it is clearly unethical. If it more like the latter case, then it may be excusable and not even considered wrongdoing. You may need to find out more information about X's motives and actions before being able to determine which case is most relevant and applicable.

Frequently, people find it easier to reach decisions regarding ethical issues in a specific case if they compare the case to similar cases that are less problematic and thus easier to evaluate. The formal term used in ethics for this kind of reasoning by analogy is *casuistry*. Casuistry is a set of practical procedures for resolving the moral problems that arise in particular real-life situations. It pays close attention to the specific details of particular moral cases and circumstances, in contrast to generalized, abstract, or hypothetical cases. Casuistry acknowledges a multitude of moral paradigms, principles, and maxims and considers circumstances as only one of many relevant moral considerations. It analyzes particular moral problems by analogy to prior paradigm cases, rather than as unique, isolated, or totally unprecedented instances.

Casuists begin by comparing particular types of cases for their similarities and differences on a practical level. From these comparisons, they develop a moral taxonomy, or detailed and methodical map of significant likenesses and differences. Casuists then proceed to reason by analogy from clear-cut cases, where the resolution of a particular ethical issue is not controversial (paradigms, or type cases), to cases that are more problematic. This approach of reasoning by analogy may be more helpful in deciding ethical dilemmas than approaches that apply pre-established principles, such as general prohibitions against cheating, lying, and stealing.

The practice of casuistry involves more than an application of established moral principles; it also requires practical wisdom, an ability to understand when and under what circumstances and conditions the rules are relevant and should apply in a particular situation, as well as how to apply them. An unresolved issue for casuists is the role, if any, of general principles.

Here is an oversimplified example of casuistical reasoning (it also appears in chapters 5 and 6). Suppose that each of three people, Ann, Barbara, and Carl, is in the same class with a roommate. In Case A, Ann copies her roommate's class notes, without permission, in order to study for a test. Assume that she had missed two weeks of class because of illness. Rather than trying to decide Case A by applying a principle such as "stealing is wrong" in order to decide that Ann's conduct was unethical, or the principle that illness is a justifiable excuse for using a classmate's notes rather than relying on one's own, casuistry uses other cases as exemplars. In this example, assume a Case B in which Barbara copies her roommate's answers on her final take-home exam, again without permission, because she had been partying too much to study properly. Also assume a Case C in which Carl and his roommate study for an exam together and Carl, with his roommate's permission, borrows notes from a class he missed. To decide Case A, a casuist will need to figure out whether it is more like Case B, which is clearly unethical, or like C, which is probably not unethical.

The casuistical approach not only helps us decide whether Dr. X's actions were right or wrong, but it can also help us decide what course of action we should take. Paradigm cases are often associated with specific resolutions; for example, if we decide that Dr. X's case is just like the time Dr. Y was caught cheating, and Dr. Y was fired, then we have at least a guide to how we should respond to Dr. X.

Although framed in terms of rules or maxims, casuistically established resolutions of cases are only general and presumptively valid, not universal or invariable. They hold good only in typical conditions, rather than in all circumstances. The partiality or tentativeness of resolutions to moral dilemmas reached using casuistry has been viewed as an asset as well as a liability of this approach. Casuistically derived resolutions may not provide the level of certainty and universal applicability desirable in resolving moral dilemmas in research ethics. Furthermore, casuistry has historically been maligned as permitting decisions that are unprincipled and discretionary. As the merits of the case analysis approach to ethics have been rediscovered, so has the debate over the practice of casuistry (Jonsen and Toulmin 1988: 239–249).

Virtue Ethics

In this approach to ethical analysis you ask: What is Dr. X's moral character? Is he generally an honest and trustworthy researcher, or has he exhibited prior instances of objectionable conduct? If I approach him with my knowledge of the errors, is he likely to deal with me honestly, or will he attempt to take advantage of my good will by attempting to cover up wrongdoing? If he is of generally reputable character, would my telling others be likely to damage his reputation and character? And what about *my* character? What are my motives in pursuing these various alternatives? Am I only interested in justice being done and having the error corrected? Or do I secretly hold a grudge against Dr. X, which this situation would enable me to act upon by damaging his reputation and credibility as a researcher through publicizing the errors as though they were intentional, when they may, in fact, not have been?

In contrast to the other approaches, which to some extent all focus on what people do rather than who people are, the ethics of virtue focuses on the character or moral quality of the relevant actors. Rather than focus on moral dilemmas, an ethics of virtue is concerned with the moral character of the actors themselves. A number of contemporary moral theorists have proposed a return to the tradition of virtue ethics. This tradition, which begins with Aristotle and is prominent in some religious ethics, has been commended by these theorists as providing a superior approach to moral dilemmas than either consequentialist or deontological ethics.

A moral virtue is an acquired habit or disposition to act morally and do what is right or praiseworthy. Virtue involves habits or skills (patterns of action) that we cultivate in ourselves. Virtues are also character traits that incline us to act in accordance with moral rules and standards. From the perspective of virtue ethics, character is more a function of one's tradition and community than of one's independent will or volition. Yet this approach is not deterministic, since it holds that moral agents have the capacity to shape and change their characters to become more virtuous.

The significant aspects of an ethical issue from a virtue perspective are the agent's character traits, dispositions, motivations, and intentions, rather than the conformity of the agent's conduct to established rules and standards or the outcome or consequences of the agent's behavior per se. Both rules and consequences, however, may be significant factors in the assessment of the agent's character and the virtuousness of the agent's actions. In addition to the prospective characteristics of virtue—dispositions, motivations, and intentions to act virtuously—conscientiousness, the often retrospective reflection on the moral goodness or ethical appropriateness of one's actions, is also an important aspect of a virtuous character.

Although using virtue and character to consider problems of research ethics may not seem obvious, some significant applications are apposite. Many professional codes of conduct stress virtues in addition to duties and ideals of conduct. Codes of professional responsibility frequently demand honesty, trustworthiness, integrity, and fair dealing from members of a profession. Virtue ethics may be particularly useful in assessing the ethical conduct of persons in situations involving clear violations of established moral norms of a particular profession. The virtuousness or good moral character of researchers is particularly important with respect to those engaged in experimentation involving animal or human subjects. Applying a virtue approach may also be particularly useful in determining appropriate sanctions for violations of established norms for conducting research. Especially where such violations are not limited to a particular instance but are part of an overall pattern of conduct, we may say that the wrongdoer lacks moral virtue or good moral character.

To take our running hypothetical, imagine attempting to determine the appropriate sanction for Dr. X's failure to report data accurately in a publication. A virtue approach might result in concluding that a lesser sanction should be imposed if he was careless and sloppy than if he deliberately misrepresented the data. Similarly, if the infraction was a first-time occurrence, we would be more inclined to say that the action was out of character than if it were a repeat offense. In the latter instance, we might characterize Dr. X as dishonest. But even in this case, determining whether Dr. X's conduct reflects a morally dishonest character will depend on the particular circumstances. The particular facts of the context in which the dilemma occurs are of central relevance to the moral evaluation in virtue ethics.

There are some obvious limitations to using a virtue approach in research ethics. One limitation is that morally good persons do not always know what is morally right, despite their conscientiousness and best intentions. Another limitation is that the individualized nature of the virtue approach may not provide objective standards that establish a "moral minimum" for morally acceptable conduct in research. Such standards are necessary to provide a way of assessing individual conduct to determine whether it measures up as acceptable. Honesty as a character trait may not be virtuous if it is used to maliciously discredit a colleague. Virtues are thus dependent upon other moral principles or ideals. As James Childress and Tom Beauchamp state, "if virtues are dispositions to act rightly, it remains important to determine right acts" (Beauchamp and Childress 1989: 264). Furthermore, claims to conscientiousness may simply be attempts to rationalize or legitimate immoral acts. Although virtues and character are indispensable to the moral life, they are not sufficient. Similarly, conscience may be mistaken or may violate clearly established moral norms or rules. Despite these limitations, virtue ethics provides an important supplement to other analytical approaches that ignore the importance of good moral character to ethical conduct in research.

Conclusion

A number of distinctive approaches to ethics may be applied to issues that arise in the conduct of research. Consequentialist approaches consider the outcomes of a particular course of action central in determining its moral rightness or wrongness, although they may also employ rules. In contrast, deontological approaches focus on duties and rights. Casuistry begins with particular facts and makes comparisons between the case at issue and other related, paradigmatic cases to determine whether the case involves ethical or unethical conduct. In contrast to all of the other approaches, virtue ethics focuses centrally on the character of moral agents and considers who they are more central to the ethical analysis than what they do.

None of these approaches provides complete or perfect guidance for resolving the variety and complexity of moral dilemmas that the researcher may encounter. Nevertheless, all of them offer guidance to researchers and others concerned with resolving ethical dilemmas encountered in the practice of research. These approaches, singly or together, have been incorporated more or less explicitly in the cases presented in this volume. In all situations, the ethical analysis of a research issue will benefit from attending to the facts, the various ways the facts can be or have been interpreted, the consequences of a particular course of conduct, the researchers' obligations to others, the rights of others, and the character of the researcher.

Note

1. A moral agent is simply some entity that is morally responsible, that can make moral and immoral decisions and take moral and immoral actions. It is a useful phrase because corporate entities (businesses, governments), not just persons, may be moral agents.

References and Suggestions for Further Reading

Almond, Brenda. "Rights." In *A Companion to Ethics*, ed. Peter Singer, pp. 259–269. Cambridge, MA: Blackwell, 1991.

Anscombe, G. Elizabeth M. "Modern Moral Philosophy." *Philosophy* 33: 1–19 (1958).

Beauchamp, Tom, and Childress, James. *Principles of Biomedical Ethics*. 3d ed. New York: Oxford University Press, 1989.

Beauchamp, Tom, and Walters, LeRoy. *Contemporary Issues in Bioethics*. 2d ed. Belmont, CA: Wadsworth, 1982.

Davis, Nancy. "Contemporary Deontology." In *A Companion to Ethics*, ed. Singer, pp. 205–218.

Frankena, William. *Ethics*. Englewood Cliffs, NJ: Prentice-Hall, 1973.

Goodin, Robert. "Utility and the Good." In *A Companion to Ethics*, ed. Singer, pp. 241–248.

Jonsen, Albert. "Of Balloons and Bicycles—or—the Relationship between Ethical Theory and Practical Judgment." *Hastings Center Report* 21(5): 14–16 (September/October 1991).

Jonsen, Albert, and Toulmin, Steven. *The Abuse of Casuistry: A History of Moral Reasoning*. Berkeley: University of California Press, 1988.

MacIntyre, Alasdair. *After Virtue*. Notre Dame, IN: University of Notre Dame Press, 1981. (Develops a neo-Aristotelian approach to virtue ethics.)

Mappes, Thomas, and Zembaty, Jane. *Biomedical Ethics*. 2d ed. New York: McGraw-Hill, 1986.

May, William. "Doing Ethics: The Bearing of Ethical Theories on Fieldwork." *Social Problems* 27: 358–370 (1980). (Discusses the relevance of teleological, utilitarian, deontological, critical philosophical, and conventional ethical theories to fieldwork.)

McGlynn, James, and Toner, Jules. *Modern Ethical Theories*. Milwaukee: Bruce, 1962. (Chapter on situation ethics provides a useful introduction to this approach.)

Pence, Greg. "Virtue Theory." In *A Companion to Ethics*, ed. Singer, pp. 249–258.

Petit, Philip. "Consequentialism." In *A Companion to Ethics*, ed. Singer, pp. 230–240.

Reece, Hayne, and Fremouw, William. "Normal and Normative Ethics in Behavioral Sciences." *American Psychologist* 39(4): 863–876 (1984). (Distinguishes the norms applicable to the practice of scientific research from the normative values relevant to ethical research.)

Rosen, Bernard, and Caplan, Arthur L. *Ethics in the Undergraduate Curriculum*. Hastings-on-Hudson, NY: Hastings Center, 1980. (A bit dated, but still useful background on teaching ethics at the undergraduate level, methodologies and substantive issues, and problems of pedagogy.)

Singer, Peter, ed. *A Companion to Ethics*. Cambridge, MA: Blackwell, 1991. (A collection of essays on a diverse range of specific topics relevant to ethical theory and practice. Those particularly applicable to research ethics are included separately in this bibliography.)

United States Department of Health and Human Services. Public Health Service. National Institutes of Health. Office for Protection from Research Risks. *Protecting Human Research Subjects: Institutional Review Board Guidebook*. Washington, DC: Government Printing Office, 1993.

United States National Commission for the Protection of Human Subjects of Biomedical and Behavioral Research. *The Belmont Report: Ethical Principles and Guidelines for the Protection of Human Subjects of Research*. Washington, DC: Government Printing Office, 1978.

Warwick, Donald P. *The Teaching of Ethics in the Social Sciences*. Hastings-on-Hudson, NY: Hastings Center, 1980. (A bit dated, but contains useful information about courses on ethics, past and present attitudes toward teaching ethics relevant to the social sciences, and discussion of topics and issues relevant to teaching about ethics in the social sciences.)

PART II

Cases in the Natural Sciences: Biology

3 The Professional Scientist

THE EIGHT CASES presented in this chapter are designed for graduate students in the biological and medical sciences, that is, for people just beginning their research careers. They are intended to illustrate the simple fact that decisions which are, at least in part, ethical are customary features of the professional life of a scientist. Even though few problematic situations have unambiguous right and wrong solutions, it is important for novice researchers to develop the habit of considering the ethical implications of decisions made by themselves and others. These cases are meant to provoke such consideration. Moreover, by discussing cases like these with their students, mature scientists can convey a crucial message: Questions of ethics cannot be left behind among the passing concerns of collegiate life; rather, they require the routine and serious attention of working professionals.

Although the cases in this chapter cover only a portion of the situations worthy of discussion, they constitute a useful beginning. Once the habit of thinking about these matters has been encouraged, we suspect—and our experience suggests—that issues and situations for discussion will suggest themselves in abundance. We feel quite confident that, once encouraged in such endeavors, individual discussion groups and individual students can carry on.

The cases concern important issues: defining professional malpractice, the response to fraud, publication rights and responsibilities, laboratory relations, and the special responsibilities of teachers. Some of these topics, such as fraud and misrepresentation of data, have attracted considerable public attention recently. Others represent important recurring themes in the relationship of scientists to one another and to society. It seemed important to us that the cases be realistic, and we hope that they are also both interesting and fun to discuss.

It may surprise some readers to find that we have not emphasized research behavior that is obviously reprehensible. If a scientist falsifies results, his or her behavior is unquestionably unethical and may also be illegal. Similarly, such practices as manufacturing data, demanding sexual favors in return for grades or letters of recommendation, and plagiarism are all plainly unethical and unlawful. It would be a rare student who could not identify them as such. With respect to such behavior, the responsibility of instructors falls mostly in the realm of reinforcing decent behavior, both by example and by ensuring that transgressors are punished. In short, we have omitted these kinds of situations because we are at a loss to imagine a two-sided discussion of fraud and because instructors can readily supply their own lists of thou-shalt-nots. We have, however, included a case dealing with appropriate ways to respond to fraud (see Case 4).

We have concerned ourselves here with the kinds of situations that arise daily in the lives of working scientists trying to behave decently while faced with numerous contradictory pressures. Since none of the cases has a simple right or wrong answer, we cannot provide an answer sheet. Instead, the problem is to identify the several obligations—some contradictory—that may bear on a scientist's decision making and to try to locate the realm of ethical behavior between extremes that are often unacceptable.

In this connection, the instructor plays a crucial role. It is common, especially among students, to suppose that decisions are of only two kinds. The first kind has a simple right or wrong answer. The second kind doesn't, and, so the reasoning goes, every opinion and every decision is equally valid. A derivative idea is that behavior is unethical only if some kind of jury could declare it to be so. This dichotomy oversimplifies life and is pernicious. If there is one critical goal of discussions using these cases, it is to reveal that there are ethical constraints even where certainty is lacking.

The model forum we have in mind for presenting these materials is the kind of intensive reading seminar that is often offered specifically for first-year students to help ease the transition to graduate school. This transition period is a good time to introduce the idea that ethical considerations are intrinsic to professional decision making and that part of the business of science is to discuss these matters.

One source of ethical ambiguity in scientific research is uncertainty about the facts. It is difficult to describe a case precisely without writing a novel. In Case 5, for example, only Big-Shot and Earnest really know the details of their collaboration, and even they may have differing perceptions. Still, any particular set of facts will limit the range of ethical options. We urge discussion leaders to play with the facts of the cases and to change the conditions, saying, for example, "OK, you don't find that bothersome, but what if the situation were as follows? . . . " Discussed in this way, these cases and their obvious derivatives can cover considerable ethical territory.

Even setting aside factual ambiguity, it may prove impossible to do more than outline a range of ethically acceptable choices. We believe that the instructor should derive satisfaction from this not inconsiderable accomplishment.

Cases 1, 2, and 3 cover various aspects of the question of when previously reported results must be corrected. They include instances in which a correction is either appropriate or absolutely necessary and instances in which one is almost certainly not required. The instructor may wish to pick and choose among these cases, since they are similar—although not so repetitive as they appear at first glance. In our experience, thorough discussion of these three cases can require two to three hours. Discussions of Case 2 tend to turn on the students' understanding of the concept of statistical significance, and the instructor can invite such discussion by electing to use the case.

Case 4 explores the response to fraud. It is an excellent opportunity to discuss the specifics of how to respond in a local context, that is, whom to contact, what the university's rules and obligations are, and so forth. Case 5 concerns the relationship of students and postdocs to a research mentor. It tends to provoke open-ended discussions. Case 6 is a two-part consideration of intellectual property rights; Part A is fairly straightforward, but Part B elicits heated discussion.

Case 7 considers one possible ethical dilemma that might confront a graduate teaching assistant. Case 8 speaks to the question of intellectual independence.

■ Case 1

Dr. Able, a microbiologist, has developed a superb new technique for introducing DNA into bacterial cells. Her technique, the "Zero Amp Penetration," or ZAP, should make life very much easier for microbiologists everywhere. Dr. Able and graduate student Baker tested the method on a wide variety of bacteria growing in many standard media, and after these extensive tests, they wrote up their results for publication: "The Gene ZAP, a New and Generally Applicable Method for Transforming Bacteria." While awaiting publication, they circulated preprints of the manuscript to a number of their colleagues who were anxious to try the ZAP.

About a month after receiving a copy, Dr. Clever called to say, "What are you people up to? I can't get the ZAP to work at all." After a little discussion, it turned out that Dr. Clever was growing her bacteria in a medium containing the amino acid tyrosine. Able and Baker had tried many standard media, but they had never tried culture media containing tyrosine. Sure enough, Baker was able to demonstrate that tyrosine completely blocks the ZAP, but he also discovered that if one adds commercial banana extract, the tyrosine block is eliminated and the ZAP works perfectly.

Part A

Suppose that while Able and Baker are working this out, their paper appears in print.

Baker is feeling a bit sheepish about what he sees as an oversight; he wants to publish an erratum or correction. Able says that's nonsense; no one ever got anywhere publishing corrections. Besides, Clever deserves credit for stimulating them to discover a more general technique. Able thinks that all three of them should publish a brief paper describing the tyrosine block and the effects of banana extract.

Clever thinks it might be inappropriate for her to be an author of the second paper. After all, her only contribution was to complain. Baker did all the work (with Able's advice). Still, Clever's grant is coming up for renewal, and she did spend a month trying to get the ZAP to work.

What is the appropriate course of action?

Part B

Now suppose that the paper hasn't yet been published when Baker discovers the wonders of banana extract.

There is still time to make changes. Able and Baker can easily add a paragraph (in the proofs) detailing the changed procedure. They might begin it, "Dr. Clever has pointed out to us . . . "

This strategy seems to solve the problem, but what about Clever's credit and, by extension, her grant?

■ Case 2

Part A

Dr. Able turned her inventive mind to novel antibiotics. Her friend, the chemist Dodd, had recently and triumphantly worked out the structure of coleopterine, a natural product from beetle dung. Dodd wondered: Does this stuff do anything? Able has a standard test she uses to look at the antimicrobial activity of chemicals, and, with this test readily available, the two of them decided to test coleopterine for antimicrobial activity, despite the fact that its peculiar chemical structure did not suggest any grounds for optimism. They were delighted to discover in their preliminary tests that coleopterine seemed very effective against pathogenic Streptococci. With these preliminary data in hand, they planned and carried out an extensive series of tests.

The results indicated that at reasonable concentrations, coleopterine is an effective antimicrobial acting specifically against Streptococci. But coleopterine was not nearly active enough to be pharmaceutically interesting; in fact, the results were of the kind often described as marginal. However, they were statistically significant (at the standard 5 percent level). Since it was surprising to find any antimicrobial activity in a compound with coleopterine's structure, and since antimicrobial specificity is always of interest, Able and Dodd published their results.

A year or so after publication, Able returned to the study of coleopterine. She started out by attempting to repeat her earlier results. Her new results were similar to the old ones, but now—when she combined all her data—it appeared that coleopterine was probably without effect; the aggregated data were insignificant at the 5 percent cutoff level.

What should Able (or she and Dodd together) do? They have several options, including:

(1) Try repeating the experiments a third time.

(2) Write off the research on coleopterine as a good try and move on to something more productive.

(3) Write and submit a brief paper reversing the conclusions of the first. (This option assumes that the paper would pass review and be accepted by a journal, an assumption that may not be justifiable.)

(4) Publish a formal retraction containing a statement that the first paper was in error.

Part B

Dr. Friendly, a professor of clinical microbiology, has also been studying a new antimicrobial. Oderiferol has been thoroughly tested in the laboratory, and it is quite clear that it is a potent antimicrobial—one that works against previously untreatable pathogens when tested on cultured bacteria or in mice, rats, and chimpanzees. Friendly is doing the first test on humans, and his results will help decide whether oderiferol can be approved for large-scale clinical tests.

Friendly's experience parallels Able's. The first test is marginally successful; he publishes; then further work reveals that the earlier results were a statistical fluke.

(1) What should Friendly do? Should Friendly's case be resolved the same way as Able's? Why or why not?

(2) What difference, if any, does it make that Friendly's research involves humans?

(3) What obligation, if any, does the journal have to publish a correction, retraction, erratum, addendum, or follow-up article? Should its obligation depend on the particular form chosen?

■ Case 3

Obviously the line between what needs to be corrected and what doesn't can be a difficult one to establish. To explore it further, we present a number of possible situations. In each instance, assume that Drs. A, B, and C have jointly published a paper in which they tested the expression of 25 different artificial genes introduced into KR3J9 cells. Each construct was tested ten times, and the results were analyzed statistically. They discovered that introduction of the dinucleotide GG at position −73 elevated transcription significantly. This was true for 20 of the 25 constructs; in the others there was no effect.

This is a conclusion of some significance for their field. Subsequently, i.e., after publication, they discover that something they wrote is untrue. The questions are: (a) Does it matter? If so, (b) what should they do about it?

(1) They discover that their program for calculating standard errors was incorrect. In every positive case the statistical significance is less than they reported, but the results are still significant at the 5 percent level.

(2) They made an error in calculating data for one construct. Only 19 of the constructs were really positive.

(3) They erred in calculating the data for the five negative constructs; the effect was really quite general, working in all 25 cases.

(4) A and B discover that C, who did the assays, always put the experimental tubes in one water bath and the controls in a second. The possibility exists that there is a slight temperature difference between the two water baths.

(5) A and B discover that C, who did the assays, always put the experimental tubes in one water bath and the controls in a second. A and B show that the whole effect is due to a temperature difference.

(6) They find that three of the constructs were not sequenced properly. They had introduced nucleotides in these cases, and the added nucleotides did boost transcription, but they weren't GG.

(7) The paper included an autoradiogram showing the effect for ten constructs. The legend said that it was a five-hour exposure. B had misread his notes; it was a 25-hour exposure.

(8) They described their transfection conditions as being "according to Schnzzlehoff, *et al.* (1987)." In fact, as A now discovers, Schnzzlehoff added a very low concentration of merthiolate to his buffers to inhibit bacterial growth. A, B, and C had omitted the merthiolate, since bacterial growth is not a problem. It is hard to imagine that the merthiolate had differential effects on results with different constructs, although it might have some general effect on cells.

(9) C discovers that a similar result, with much less data, had been noticed by Tattenheimer in 1986 and ignored by everyone. Since they hadn't remembered Tattenheimer's paper, they didn't cite it.

(10) They had done two kinds of assays: (a) C did enzyme assays; (b) B did RNA measurements. Now they discover that because of systematic error, all the enzyme assay data can't be trusted. Alternatively, suppose A and B discover that C lied about the enzyme data.

■ Case 4

Part A

Armstrong is a first-year graduate student in a molecular biology laboratory, and Hayes is her scientific idol. Hayes, who is just finishing his thesis work, has a golden touch. Every experiment he does produces clean, crisp data: the lines fall through the points, scatter is less than or equal to theoretical predictions, etc. Because his experiments seldom need to be repeated, Hayes has produced an awesome thesis, full of new, fascinating, and demonstrably correct results. Indeed, the laboratory has already followed up on many of these leads with telling success.

One day Armstrong notices Hayes leaving the scintillation counter and can't help noticing that he has 80 vials. This barely registers in her subconscious until later in the day when he shows her the—typically beautiful—experimental results with 40 data points. When she asks about the missing points, he explains that it is standard practice to eliminate outliers from the analysis. "What!" Armstrong says, shocked, "50 percent of your points were no good?" "Mind your own business," Hayes replies sweetly, going about his.

Armstrong is a bit distraught. But at last she summons up her courage and tells their professor this story. He can barely suppress his boredom and his annoyance—he had expected her to present him with results from experiments she has been procrastinating about for two months.

(1) Is this really Armstrong's business? If so, what can she do?

(2) Suppose you are Jones, the department chair. After failing with the professor, Armstrong comes to you. What can you do?

(3) Suppose that you are an interested reader of this work. The data strike you as too good to be true, and you write to the professor asking to see copies of the original data (counter tapes and lab notebook pages). You think there may be an error of interpretation. Does the professor have a responsibility to give these materials to you?

(4) Suppose as chair you make the same request, saying you have reason to suspect fraud on Hayes's part. Does the professor have any responsibility to show them to you?

(5) Let's suppose that the chair succeeds in proving the fraud before anything has been published. Hayes agrees to leave graduate school, and because he doesn't force the issue, no formal hearing is convened. (No one wants to do the paperwork.) Nothing more is said. Are there any problems with this solution?

Part B

Let's suppose that Armstrong does nothing, and Hayes's work eventually fails. After publication, other labs try hard to repeat it, and despite the original impression that it was all demonstrably correct, no one can get things to work as he did. After an unimpressive postdoc, Hayes has gone on to a good job as a stockbroker, and the professor has moved on to other research topics. Now Armstrong mentions her story to another member of the faculty. The story suggests that the problem wasn't simply unavoidable error, but fraud. Does anyone have any responsibility at this point?

■ Case 5

Part A

Dr. Big-Shot's postdoctoral associate, Dr. Earnest, has isolated a novel gene product (protein C), which may lead to a cure for the common cold. When Earnest leaves, she will wish to continue studying protein C. But Dr. Big-Shot wants to work on the same project, since it is the logical extension of his lab's work and since his grant funded the research. Funding agencies are very reluctant to approve two grants to conduct precisely the same research.

(1) If Big-Shot and Earnest cannot agree on a collaboration, who should get priority?

(2) Would it matter to your analysis if Dr. Earnest had been supported during her stay in Big-Shot's lab by an independent postdoctoral fellowship?

(3) Would it matter if Earnest's results were not in the main line of research in Big-Shot's lab?

Part B

Many journals demand that, after publication, any unique research materials (mutant stocks, clones, etc.) must be provided to any academic research worker who requests them. Dr. Hotshot is Dr. Big-Shot's main competitor. Big-Shot is afraid that if he and Earnest publish too soon, Hotshot—who will then have a right to their materials—will be first, will scoop the patents, and will take the Nobel prize. But if they delay publication, Dr. Earnest will have a much more difficult time getting a job and applying for grants.

(1) Is it legitimate for Dr. Earnest to submit the manuscript without Dr. Big-Shot's approval?

(2) What if Dr. Earnest agrees to wait for further results, but five years later Dr. Big-Shot still refuses to publish?

Part C

Assume Drs. Earnest and Big-Shot have agreed on a collaboration, so that each will work on one aspect of this very hot project. Now a beginning faculty member, Dr. Earnest must apply for her first grant, prepare courses, learn how to serve on committees, attract graduate students, and wait ten months for the university to finish renovating her laboratory. All the while, this collaborative project is waiting.

(1) How long should Dr. Big-Shot wait before assigning that portion of the work that has been granted to Dr. Earnest to one of the ten eager postdocs in his lab?

(2) What about the rights of the public? How are they best served? What if the gene might identify a cure for AIDS instead of the common cold?

Part D

Jones, a graduate student in Dr. Big-Shot's laboratory, works out a procedure for large-scale production of protein C. The critical element is a particular plasmid that Jones has constructed. Then Jones accepts a job at a major biotechnology company.

(1) Should Jones be able to take his plasmid of the common cold with him to the biotech firm? How about his notebooks?

(2) Would your answer be different if it were Dr. Big-Shot who made the contract with industry?

(3) In industrial contexts, individuals who switch jobs between competitors may face a court injunction forbidding them to share particular information for a certain period, particularly if they have signed a no-competition agreement with the first employer. Should such a restriction be applicable to Jones?

Part E

Eventually protein C becomes a commercial success. Drs. Big-Shot and Earnest, and their universities, all get a share of the profits. What, if anything, is owed to other individuals who worked on this project in Big-Shot's laboratory before Dr. Earnest took the last step and found the gene for protein C?

■ Case 6

Part A

Dr. Andrews and Dr. Barr are friendly competitors working on ways to predict protein structure. Barr discovers a simple mathematical algorithm for accurately predicting the locations of domain boundaries in proteins, and at an early stage of testing his ideas, he begins to discuss them at meetings. Andrews is very skeptical at first, and they have a number of heated discussions at meetings, but at the same time he tests the ideas himself. Much to his surprise, he finds that the algorithm is terrific. He devises 20 tests (quite different from Barr's). When the scheme passes all of them, he writes the story up for publication.

(1) Given that Barr hasn't yet published, does Andrews have any obligation to cite his work?

(2) What if Barr hadn't talked about the work at meetings, but the two had talked informally? Alternatively, what if the work had been presented in a poster?

(3) What if Andrews had read about Barr's work in a grant proposal?

Part B

You operate a very active lab investigating, let us say, regulation of the gene X by the retinoblastoma protein. One of your competitors is John Smith; he works on regulation of gene Y by retinoblastoma, and your relations with him are amicable, though not great.

Both of you have reasons to think that retinoblastoma protein normally modulates expression of the genes you study, but—as you have published previously—when the retinoblastoma protein is overexpressed in cells (by transfection), it has little effect on expression of the X gene. You are aware that it might be worthwhile to look at the effect of transfected retinoblastoma on a transfected X gene, and you have written yourself a note to that effect, but—since it's highly unlikely that the results will be different—the experiment is not a high priority for you.

Out of the blue, you receive a manuscript for review from the journal *Cell*. The senior author is Smith; he and his colleagues are reporting that transfected retinoblastoma product regulates transfected gene Y. Objectively, the manuscript is poor: most of the data are extremely sloppy, and some experiments are improperly controlled. Your task as a reviewer is, therefore, a straightforward one; the paper is simply not appropriate for *Cell* in its present form.

Your prompt review enumerates the manuscript's flaws and demands the appropriate control experiments. But it points out that the root observation may be true (although the data are unpersuasive) and that, if true, it is important. Subsequently, the editor faxes you his thanks, indicating that the manuscript has been rejected pending major revisions.

Although the manuscript is out of your hands, its fundamental observation is, if true, obviously of some importance to you, so you do the obvious thing: you try the experiment using gene X. If the experiment fails, there is, of course, little to talk about. But let's suppose that you see a dramatic effect. Let's suppose further that your skills are so impressive that within days you have obtained enough data concerning this effect to warrant publication in a first-rate journal.

There is a problem here. What should you do with your data? Some of the following questions may be worth considering:

(1) Do your relations with Smith matter?

(2) Does it matter that you have previously written yourself a note to do this experiment?

(3) Should you have refused to review the paper?

(4) How do you avoid having the nature of your review affected by the competition?

(5) Is there anything wrong with trying the experiment yourself?

(6) If the experiment fails, should that fact enter into your consideration when you re-review Smith's manuscript?

(7) Should you consider joint publication with Smith?

(8) Suppose that you are in the midst of writing a grant renewal. The line of work you're trying to defend will be very much more powerful if you can note that this transfection assay works, and all you want to do is to insert a line to this effect. Should you?

In all these matters, try to distinguish among courses of action that are legitimate and those that are illegitimate—and among those things that it might be nice to do and those that you must or must not do.

■ Case 7

Snyder is an assistant instructor in a general biology course. The course has been covering introductory genetics as well as immunology. In an exciting and time-tested laboratory exercise just before the winter break, each student obtains a drop of his or her own blood for experiments, then determines his or her ABO blood type and gives the instructor a sample of white cells to be cultured for karyotyping later.

(1) When the students return after the break, Snyder is confronted by an obviously upset student who confides, "My blood is AB, but both my parents have blood donor cards that say 'Type A.' If I understand the genetics correctly, they can't be my parents!" What should Snyder say?

(2) The cultured white cells are used in the next laboratory where students determine their own karyotypes. Poor Snyder is having a bad year, for up comes an obviously terrified young woman, who asks, "Why do I have an XY karyotype?" What does Snyder say?

The following may be relevant to your thinking: (a) The ABO blood test, when carefully executed, is definitive and is considered so in a court of law. In those circumstances, the student's interpretation is correct. However, mistakes can be made in its execution and, with somewhat greater difficulty, its interpretation. Finally, blood bank records are occasionally wrong, so that one or both of the donor cards may be incorrect. (b) The XY karyotype is possible and not astronomically rare: it may indicate one or another form of male pseudohermaphroditism—a condition that is usually but not always diagnosed in adolescence. The condition calls for medical treatment and psychological adjustments. Remember, however, that neither the student nor, for that matter, Snyder is a trained cytogeneticist; Snyder probably learned how to do the karyotypes just last week—and he certainly isn't a genetic counselor.

■ Case 8

Dr. Lee is an entomologist. After receiving his Ph.D. from a world-renowned university with a much-praised thesis on the taxonomy of silk moths of the family Saturniidae, he has accepted a faculty position at a large research university. Now for the first time he is facing the fact that research funds to support taxonomic studies are in short supply. The standard agencies (NIH, NSF, USDA, ACS, etc.) have little interest in Dr. Lee's work.

However, the ecology-oriented lobbying group Society of Citizens for a Nicer Planet (SCNP) is interested in supporting Dr. Lee's work. It proposes that he accept $7,000 in support of research on the effects of malathion on the diversity of Saturniid moths. Their conditions are (a) that his results be published first as a feature article in the Society's mass-circulation *Journal of a Nice Place* and (b) that he turn over editorial control to the journal. Only following such publication can he submit his results to a professional journal.

Dr. Lee is, in fact, interested in the question of malathion's effects, and he sorely needs the research support. Do you have any reservations about his accepting SCNP's offer? If so, why?

4 Scientific Misconduct
What Is It and How Is It Investigated?

THIS CHAPTER PRESENTS a case whose theme is scientific misconduct. What do we mean when we talk about misconduct? While the scientific community has not reached a consensus on its definition, the current federal regulations state that "misconduct in science means fabrication, falsification, plagiarism, or other practices that seriously deviate from those that are commonly accepted within the scientific community for proposing, conducting, or reporting research. It does not include honest error or honest differences in interpretations or judgements of data" (U.S. Department of Health and Human Services 1989: 32449). This definition is being revised; a panel of scientists assembled by the National Academy of Sciences has recommended that it be changed to include only fabrication, falsification, and plagiarism.

The scientists also have recommended that another category, which they suggest calling "questionable research practices," be used to describe unethical and sloppy scientific conduct that is not fabrication, falsification, or plagiarism. Some of the things they would include in this category are "failing to retain significant research data for a reasonable period; maintaining inadequate research records, especially for results that are published or that are relied on by others; conferring or requesting authorship on the basis of a specialized service or contribution that is not significantly related to the research reported in the paper; refusing to give peers reasonable access to unique research materials or data that support published papers; . . . and inadequately supervising research subordinates or exploiting them" (National Academy of Sciences 1992: 28). The panel also noted that "there is at present neither broad agreement as to the seriousness of these actions nor any consensus on standards for behavior in such matters."

Federal regulations also govern how an institution receiving funding from the Public Health Service (e.g., grant funds from the National Institutes of Health) is to carry out an investigation of alleged or suspected misconduct in science (U.S. Department of Health and Human Services 1989). You may wish to consult these documents as well as the procedures in place at your own institution for discussion of this case.

This hypothetical case is loosely based on several scientific misconduct cases that have occurred in the last few years. It does not involve a whistleblower and therefore does not involve such issues as whistleblower victimization, officials failing to heed warnings and take action, or the need to ensure accusers' confidentiality during an investigation while providing due process to the accused. While these are important issues which can be explored if time allows, the intended focus here is on three equally important topics:

1. Proper laboratory practices, particularly in the management of data and its manipulation as manuscripts are prepared.
2. The definition and identification of scientific misconduct.
3. The way in which investigations of alleged scientific misconduct are and should be carried out.

Part A (One Day in August)

David Dunbar, one of the postdocs in Professor Steve Grey's lab at Big Tech, has just finished presenting the results of his latest set of experiments to the lab group at the weekly lab meeting. Grey's lab is a large group, with 22 technicians, research associates, graduate students, and postdocs all working on the identification and mechanism of action of genes associated with cancer. Dunbar presented the results of a series of experiments investigating the expression of the *tnc* cancer gene in normal cells and a variety of cancer cell lines. (The *tnc* gene is associated with toenail cuticle cancer, a rare cancer usually seen only in certain at-risk families.) It was a nice presentation. For instance, his table of RNA levels was beautiful, as well it should be. It was the last figure in a manuscript he and Grey just submitted to the *Prestigious Cancer Journal*, Dunbar's fourth paper on which he was first author since joining the lab two and a half years ago.

Shortly after the meeting, Erik Larson, a graduate student, comes into Professor Grey's office, shutting the door behind him. "David couldn't possibly have done all the analyses he reported in group meeting," asserts Larson angrily. "Unless he's got a lab at home. In fact, I'll bet some of the cell lines he showed numbers for in that last table haven't even grown up enough yet for RNA isolation." "Have we ever had reason to doubt David's work?" asks Grey. "No," Grey continues, not waiting for Larson's response. "We have always been able to follow up on his results, to the benefit of many in the lab. Yes, I know he gets a lot done during his time in lab, but he's just more experienced and better organized than most others. Now, how is that work of yours going on the search for *tnc*-related genes in yeast?"

After a short discussion of his recent experimental results, Larson leaves Grey's office. Grey sits at his desk, reflecting on their conversation. "I guess Dunbar's success just makes others uncomfortable," concludes Grey, wishing that personnel management had been part of his training.

Questions for Discussion

1. Was Larson right to bring his concerns to Grey? Could Larson have presented his concerns in a better, more persuasive way? Should Larson do anything further now that he has spoken to Grey?

2. Was Grey's response to Larson appropriate? Is there some way in which you think it could be improved?

Part B (One Month Later)

Jeff Adams, a new postdoc who has just arrived in Grey's lab, pulls Larson aside. Adams is supposed to pick up the work on human *tnc*, since Dunbar will be leaving soon. "Hey, what's with these papers you guys have published?" Adams asks, waving a paper of which Dunbar, Larson, and Grey were authors. "What do you mean?" Larson responds. "Well, look at this autorad in figure 1. All the lanes are the same!" fumes Adams. "Sure," replies Larson, "that's the point of the paper. We see the same, odd rearrangement to give a new 7.2 kb band in all the cell lines from toenail cuticle tumors." "No, that's not what I mean," says Adams, shaking his head. "Look at the background dots on the film. The same dots are in each lane. These aren't the results from different tumor lines; these are copies of the same photograph!" "Oh, my heavens!" exclaims Larson. "I never noticed that before. I was a new student in the lab when this work was done and all I did was help on the cell growth and DNA isolations. I have no idea how Dunbar made this figure. We'd better talk to Steve right away."

After talking with Adams and Larson and seeing the published figure in a whole new way, Grey calls in Dunbar and angrily accuses him of fabricating data. Dunbar appears genuinely shocked. "I didn't make up any data!" he asserts. "I did all those analyses and got those results; I can show you the autorads. I was just trying to make the nicest-looking figure for publication." "But these aren't the results for the cell lines indicated in the figure. It's all copies of *one* of them. You can't do that," replies Grey. "Why not? It's the same thing as cutting up the autorad to make figure 4. I didn't try to deceive anyone," says Dunbar. "And besides, no one said anything. Not people in the lab, not you, not the reviewers. I thought that was how it was done."

Questions for Discussion

1. Below are drawings of the two figures from the published paper (figures 1 and 4), as well as the two original autoradiograms from which they were derived (figures 2 and 3). Did Dunbar fabricate data in his production of figure 4? In the production of figure 1? Explain your criteria for determining fabrication.

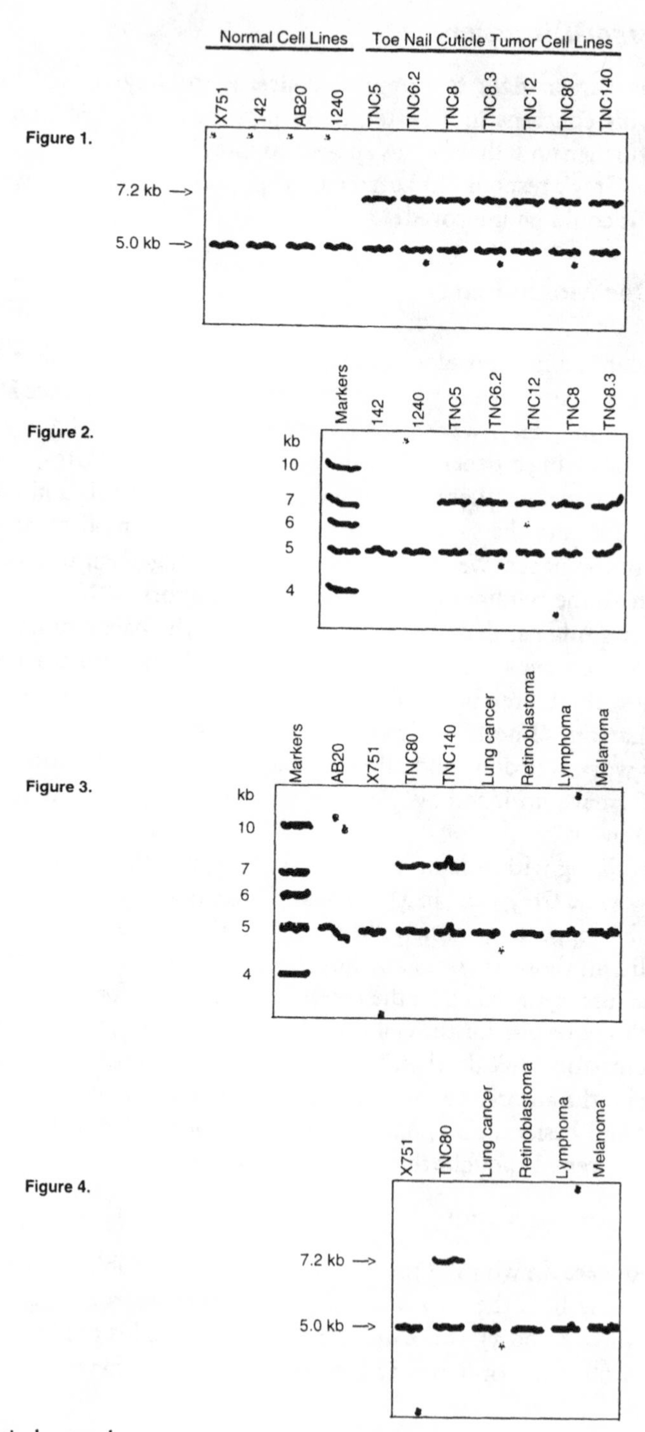

Hypothetical autorads.

2. Does it matter that the primary data show that the results are the same as shown in figure 1? Does it matter that the avowed intent was to produce a prettier picture, not to deceive?

3. Should Larson have been a coauthor on the paper if all he contributed were some routine laboratory manipulations?

4. What are the responsibilities of coauthors for the authenticity of the contents of a paper? What are the responsibilities of reviewers in this area?

5. How could Grey change the practices in his laboratory to minimize the possible recurrence of a problem of this sort?

Part C

Although somewhat relieved that the autorads Dunbar produced verified that his creative graphic artistry did not alter the basic results or conclusions of the published paper, Grey is still shaken and worried. He decides to check all of the work Dunbar did while in the lab. Grey asks three senior postdocs in the lab, Xavier, Yates, and Zimm, to begin a review of Dunbar's notebooks and published work. Grey then heads for the department chair's office to inform him of what has been discovered and what is being done.

Questions for Discussion

1. Should Grey have contacted the departmental chair at this point?

2. Are there any other interested parties who should be informed?

Part D

Dunbar returns to the lab after lunch to discover Xavier, Yates, and Zimm looking through his notebooks. Dunbar is furious, asserting that notebooks are, like diaries, private. "I thought I made it clear to the lab last year, when that new student was pawing through my notebooks, that no one was to touch them without my permission," says Dunbar.

Yates can't believe what he's hearing. "It was Steve who asked us to check over your work," Yates says, "and I think you know why. Besides, where do you get these ideas about notebooks? When I was a grad student my Ph.D. advisor routinely checked each student's notebook every evening, and anyone in the lab was free to look up any information needed for their work."

"Actually, it's a good thing you're back. We have a couple questions," interjects Zimm. "Where are the data for the first paper you published from this lab? We can only find this box of autorads."

"I threw those notebooks out a few months ago. I figured the work was all published long ago, and I needed more space on my bookshelves," replies Dunbar.

"Well, then, what about the instrument printouts for the analyses you presented in the second paper? We can only find tables recording mean values. There doesn't seem to be any record of the actual, individual determinations," says Xavier, jumping into the discussion.

"I never keep instrument printouts," replies Dunbar, getting angrier by the second. "Who does? I don't need all that useless paper cluttering up my desk. I just analyze the data and ditch that stuff. I've got better things to do than play filing clerk."

Questions for Discussion

1. To whom do laboratory notebooks belong? The individual? The principal investigator (PI)? The department? The laboratory? The university? The funding agency?
2. Who should have access to laboratory notebooks and other experimental data? Are there only certain circumstances under which some people should have access?
3. What types of data should be retained, in what form, and for how long? Whose responsibility is it to see that data are appropriately retained?
4. Where should the data be retained? For instance, do the notebooks go with a finishing student or should they stay in the lab?

Part E

Larson, rather shaken by the revelations of the day, wonders about the effect that this business will have on his career. He was so pleased to have his name on a paper published in his first year in the lab; now he's not so sure it will be to his advantage at all. In the cell culture facility, Larson, remembering Dunbar's lab meeting last month, decides to take a look at the culture logs and compare them with the lines listed in Dunbar's RNA level table in the submitted manuscript. Unfortunately, his unease was justified. Two of the lines listed in the table were not even in the lab at the time of the meeting. They arrived since then from the stock center and are being grown, but Dunbar couldn't have obtained data from them when he said he did. Larson makes a copy of the log and goes to look for Grey.

When Grey and Larson confront Dunbar with the cell culture log the next day, Dunbar admits that the numbers reported in the table were not derived from RNA analyses but were his best estimates of what the results would be. "Look," says Dunbar, "I knew what the results would be. You know how long it takes to go through the review process for the *Prestigious Cancer Journal.* It's been more than a month, and we haven't heard a thing. If I had waited for the cells to come in and grow up *and* for the review, it would have been a year, and I could have gotten scooped! Don't worry, I didn't do anything wrong.

When the paper was accepted pending a few requested revisions, I planned to just put the real data in the table and everything would be fine."

"Sure," says Larson. "How can I believe that you really would put in the real data?"

"Easily," replies Dunbar, "because that's just what I've done before."

"What! When else have you submitted 'estimated data' for review?" asks an astonished Grey.

"Lots of times. Like in that first paper with the figure you're so upset about. What's the big deal? I've never tried to deceive anyone. I've never had to change the conclusions of a paper."

Grey searches for and finally finds a disk with the file of the two-year-old manuscript (in the form in which it was submitted), calls it up on his computer, and compares it with the reprint he keeps in his top desk drawer. Sure enough, the numbers listed in the table show the same basic trend in the two versions, but are not the same.

Questions for Discussion

1. Does Dunbar's method of preparing his manuscripts for publication constitute fabrication or falsification? Is it, rather, a questionable research practice? Or is it simply a novel way to speed the progress of science?
2. Is it important to consider that the conclusions drawn in the submitted and final versions are the same?
3. Is it important to consider that, as Dunbar asserts, he did not intend to deceive anyone?
4. Is it important to consider that, in the end, only the real data were actually published?
5. What are the consequences to science of this approach to publishing?
6. Suppose that, in his fury, Larson threatens to call the editor of the *Prestigious Cancer Journal* and tell all. Should he do this? Is it warranted and proper, is it premature, or is it unwarranted and inappropriate?

Part F

While all this was going on in the Grey lab, the departmental chairman, Jack Washington, was also busy. After Grey informed him of the problematic figure, Washington consulted with the dean for research at Big Tech to see what was expected of him. He was told that he was to see that an initial inquiry was undertaken to determine whether an investigation was in order. It was up to Washington to appoint those who would conduct the inquiry. Above all, the dean cautioned, keep this quiet.

"Well, if all we have to do is gather information while keeping a lid on this, the people Grey has got looking into it will make a perfect inquiry committee," he decides. "No one else need be involved." So Washington calls Grey to request that Grey send him a written report when Xavier, Yates, and Zimm finish.

Within three weeks, the three postdocs give Grey a written summary of their findings which mentions the figure produced by creative graphic artistry and the "estimated data" submitted for review, as well as the missing laboratory notebooks and primary data. Grey reads it over, edits it a bit, and sends it to Washington. Washington then forwards it to the dean for research as the report of the inquiry committee he was told to appoint.

Questions for Discussion

1. Are other members of the same lab the best people to review Dunbar's work? If not, who would be a better choice and why?

2. Should the head of the laboratory, who is also a coauthor, be involved in the initial inquiry in the manner described here? What arguments for and against his involvement can you make?

3. Has Washington fulfilled his obligations to the institution and the accused?

Part G

After reviewing the inquiry report, the dean for research and other administrators at Big Tech decide that a full misconduct investigation of Dunbar is required. They further conclude that no investigation need be carried out for Grey or the other coauthors, as the suspect conduct seems to be Dunbar's alone. They so inform Dunbar, Grey, Washington, and the National Institutes of Health (NIH) which funded this research.

Grey contemplates what he should do. Concluding that the best way is full disclosure and a clean break with Dunbar, he dismisses Dunbar from the laboratory and terminates his salary, which had been drawn from an NIH grant awarded to Grey. Then, after consulting with the other coauthors but not with Dunbar, Grey writes to the journals retracting all the published papers on which Dunbar was an author and withdrawing the manuscript still in review.

Questions for Discussion

1. Do you also conclude that an investigation of Dunbar is warranted? If so, what would be the component(s) of scientific misconduct of which you would accuse Dunbar?

2. Do you conclude as well that only Dunbar should be subject to a misconduct investigation?

3. Are Grey's actions proper and warranted? Which, if any, are inappropriate and why?

Part H

All of this couldn't have happened at a worse time for Dunbar. After a series of successful interviews, he was looking forward to starting his own lab at another university. With the research he had done in Grey's lab and previous publications from his Ph.D. research, Dunbar figured he had a pretty good shot at a good job. Now getting a good job looked impossible because, in addition to everything that had already happened, Grey sent letters to each of the universities to which Dunbar applied telling them of the accusations against Dunbar and the planned investigation and retracting what had been very strong letters of recommendation. In response, the one university that had already offered Dunbar a position withdrew its offer.

Questions for Discussion

1. At the time the concerns about Dunbar's work were raised, Grey had already sent letters of recommendation in support of Dunbar's job applications. Was he under any obligation to inform the institutions to which Dunbar had applied of changes in his evaluation of Dunbar since writing his letters?

2. Suppose Grey has not yet written his recommendation letters when these matters come to light. Is he under any obligation to inform potential employers of the pending investigation?

Part I

The dean for research appointed a five-member committee to conduct the investigation. All five members were from the Biological Sciences Division of Big Tech. They reviewed the evidence and interviewed people at Big Tech and other universities.

When Dunbar was interviewed, he did not deny any of the actions of which he was accused, but he did deny that he was guilty of scientific misconduct. He asserted that it was never his intent to deceive and that all of the data presented in his papers were derived from actual primary data. He denied ever fabricating anything.

"I thought that was how you prepared a paper for publication," Dunbar said. "No one told me any differently. In fact, the first manuscript I ever prepared was when I came to Grey's laboratory. When I was a graduate student at Enormous State University, my advisor wrote all the papers that came out of the lab. Yes, some of my original data are gone. I didn't know that I was expected to keep them even after they were published, or that people thought instrument printouts were important. So I'm a poor document clerk; that's no crime!"

Questions for Discussion

1. Was the composition of the investigating committee appropriate?
2. Should naïveté be an adequate defense in a situation like this?
3. How can the scientific community ensure that others in the future will not also be able to say, "But I didn't know."

Part J

The Big Tech investigating committee concluded that Dunbar was guilty of scientific misconduct, having found the multicopied lanes on the autoradiogram and the submission of "estimated data" for review to be examples of fabrication. In addition, they concluded that Dunbar had engaged in many questionable research practices, such as prematurely destroying data, failing to record or keep primary data, and denying other scientists access to his data.

The findings of the investigating committee then went to Big Tech's administration for action. As required, a report was sent to NIH, but no further action was taken to punish Dunbar because he was no longer associated with Big Tech and was no longer engaged in scientific research. When last contacted, Dunbar had enrolled in an MBA program and was trying to put his life together again.

Questions for Discussion

1. Is this an appropriate conclusion for this tale?
2. What, if anything, could and should Big Tech or NIH do to punish Dunbar?

References

National Academy of Sciences. *Responsible Science: Ensuring the Integrity of the Research Process.* Washington, DC: National Academy Press, 1992.

U.S. Department of Health and Human Services. Public Health Service. "Responsibilities of Awardee and Applicant Institutions for Dealing with and Reporting Possible Misconduct in Science." *Federal Register* 54(151): 32446–32451 (August 8, 1989).

5 Authorship and the Use of Scientific Data

THE CASE STUDIES in this chapter are designed to provide the opportunity to practice applying ethical considerations to scientific research and, more important, to stimulate consideration and discussion of questions of authorship and the use of scientific data.

People frequently find it easier to come to decisions regarding ethical issues in a specific case if they compare it to other, similar cases that are less problematic and thus easier to evaluate. The formal term used in ethics for this kind of reasoning by analogy is *casuistry* (see chapter 2). The purpose of casuistry is to use clear-cut or paradigmatic cases as guidelines for resolving more difficult cases that are not as clear.

To use an oversimplified example, suppose that each of three people, Ann, Barbara, and Carl, is in the same class with a roommate. In Case A, Ann copies her roommate's class notes, without permission, in order to study for a test. Assume that she had missed two weeks of class because of illness. Rather than trying to decide Case A by applying a principle like "stealing is wrong" in order to decide that Ann's conduct was unethical, or the principle that illness is a justifiable excuse for using a classmate's notes rather than relying on one's own, casuistry uses other cases as exemplars. In this example, assume a Case B in which Barbara copies her roommate's answers on her final take-home exam, again without permission, because she had been partying too much to study properly. Also assume a Case C in which Carl and his roommate study for an exam together and, in the process, Carl borrows notes from a class he missed, with his roommate's permission. To decide Case A, one needs to figure out whether it is more like Case B, which is clearly unethical, or like C, which is not unethical.

This approach of reasoning by analogy is sometimes more helpful in deciding ethical problems than approaches that apply established principles, such as general prohibitions against cheating, lying, and stealing. In answering the questions to the following cases, try, whenever possible, to use casuistry rather than directly appealing to basic ethical principles.

■ Case 1

Rather than doing fieldwork in accordance with the terms of the six-month grant he received to work in Hawaii, Albert Atos, a graduate student, has played the tourist and enjoyed the sights of the islands. However, his failure to do the field research has not prevented him from writing a research paper on the behavior of tropical fish. He has simply copied, with slight modification, a study conducted and written five years ago by a student at the University of Singapore that was never published.

Questions for Discussion

Are Atos's actions ethical? Why or why not?

■ Case 2

Belinda Bishara, another graduate student, has been part of a large laboratory's efforts over the past five years to identify the gene that causes cystic fibrosis. She is one of ten graduate students, technicians, and postdoctoral fellows who have worked closely together on this project, doing a large number of interrelated experiments. In writing a report on her research, she has included, with permission and citation, some unpublished data from other members of the laboratory so that the context of the experiments she performed would be clear.

Questions for Discussion

Are Bishara's actions ethical? Why or why not?

■ Case 3

Chad Chan didn't attend his introductory biology laboratory section last week. Now it is 2:00 A.M. and the laboratory report is due at 9:05 A.M. Chan is desperate because a poor grade in this course would put him on academic probation, so he pulls out the report his roommate wrote last year for this unit of the course. Chan's roommate gave Chan all of his notes and reports at the start of the term to help him get through the course. So Chan decides to copy his roommate's old data and report, changing a phrase here and there, and hand it in.

Questions for Discussion

1. Is this case more like Case 1 or Case 2?
2. Do you conclude that Chan's actions are ethical? Why or why not?

Case 4

A second undergraduate student in the same biology course, Donald Dunn, is a chronic klutz. Everything he touches either breaks or does something odd. This is especially true in science laboratory courses. So Dunn and his lab partner have arrived at a mutually satisfying solution: Dunn prepares the pre-lab protocol to guide their work and then records all the data generated in lab. His partner does all the hands-on experimental manipulations. They then take the data and independently use them to prepare laboratory reports.

Questions for Discussion

1. Is this case more like Case 1 or Case 2?
2. Do you conclude that Dunn's actions are ethical? Why or why not?

Case 5

Since Dunn's laboratory section has an odd number of students, Elaine Estoban has chosen to work on her own rather than in a group of three. She works very hard during lab periods and has done well in the course. One evening while reviewing her data, she discovers that they make no sense. There is no line generated when the results are plotted, and Estoban is at a loss as to what to do. She calls Dunn to ask for advice. As far as they can tell, Estoban has been plotting the data properly and understands the principles the experiment was designed to demonstrate. Dunn says that Estoban should take his own data set, modify it a bit, and then use that to write her report. Dunn dictates his data to Estoban over the phone, and Estoban then does as Dunn suggests to prepare her report.

Questions for Discussion

1. Which of the preceding cases do you consider to be most similar to this one? Are there significant differences?
2. Are Estoban's actions ethical? Why or why not?
3. Are Dunn's actions ethical? Why or why not?
4. Rank the seriousness of the ethical issues raised in Cases 3, 4, and 5. If you were the instructor for this biology course and discovered what your students were doing, what action would you take in each case (for instance, do nothing, talk to the student, give the student a zero on the particular report involved, fail the student in the course, or report the students involved to the campus student ethics committee)?

■ Case 6

Professor Francine Field is planning to do an experiment to test whether plant growth is influenced by music. A friend has just sent her a set of old issues of a Hungarian botany journal knowing that she likes to collect old journals. On paging through one of these volumes (she happens to read Hungarian), Professor Field discovers that the experiment she plans to do has already been done and was published ten years ago by another researcher who has since died. No one in the U.S. knows of this work, and the results are those expected by Professor Field. She tells herself that there is no point in wasting money repeating the research and that botanical science needs to know these results. Besides, she feels that she should receive credit for her idea. So she takes just the Hungarian data, which she presents as her own, and writes a new paper in which she draws conclusions somewhat different from those in the Hungarian article. Upon publication, her peers are impressed with her work, and she receives several invitations to present seminars.

Questions for Discussion

1. How is this case similar to Case 1? To Case 4?
2. If there are significant differences between this case and Cases 1 and 4, what are they?

■ Case 7

Gail Ghana, a graduate student in Professor Helen Hiromoto's laboratory, recently completed a series of ten experiments designed to test a model proposed by Professor Hiromoto. Professor Hiromoto originally proposed the model to explain an unpublished experimental result generated by Janna Johanson, a former graduate student in the laboratory. Ghana wrote up her results and submitted the manuscript for publication.

The summary received from the editor indicates that the reviewers were pleased with the quality of the paper and its importance to the field, but they have recommended that one additional experiment be included before the paper can be accepted for publication. This additional experiment is the unpublished work of Johanson.

Unfortunately, Johanson left the laboratory after only one unpleasant year. She and Professor Hiromoto never seemed able to be civil to each other. Johanson finally left the lab to go to medical school, vowing that she would do everything in her power to make life difficult for Hiromoto. Certain that Johanson will refuse to have her work included, Hiromoto and Ghana do not contact her, and Ghana easily generates the table requested by the reviewers using the relevant data from Johanson's notebook (which remained in the lab, like all notebooks of Professor Hiromoto's students). Hiromoto feels that the data are hers to control and use, since the work was done in her lab, under her direction, and using money from her grant. Professor Hiromoto decides not to cite Johanson, fearing trouble.

Questions for Discussion

1. Do you consider this case to be similar to any of the previous cases? If so, which one(s) and why?

2. Which, if any, of the actions taken by Ghana and Hiromoto were improper?

3. Consider the actions of the people described in all the cases of this set. Placing Case 1 at one end and Case 2 at the other, where would you position the other cases along this continuum from clear appropriation of another's work and ideas (Case 1) to proper permission and citation (Case 2)? Be prepared to explain your reasons for your relative ranking of the seriousness of the ethical breach(es) in each case.

6 Data Alteration in Scientific Research

THE CASE STUDIES in this chapter, like those in chapter 5, are designed to provide the opportunity to practice applying ethical considerations to scientific research and, more important, to stimulate consideration and discussion of data alteration in scientific research. And like chapter 5, this chapter uses a casuistic, or reasoning-by-analogy, approach (see chapters 2 and 5 for a description of this method). In answering the questions to the cases, try to use casuistry whenever possible rather than directly appealing to basic ethical principles.

■ Case 1

Adam Able and Bob Baker are lab partners in an introductory biology course. Both are determined to attend medical school and thus believe that they must receive A's in all of their science courses. They have developed a unique strategy to do well in this biology laboratory. For each session, they arrive well prepared and well versed in the experiment to be carried out, and they work hard and efficiently in the lab, noting their results in their notebooks. However, each is actually keeping two sets of notebooks. In the evening, Able and Baker get together and create a data set for the experiment that will be very close to that expected if the experiment had worked perfectly. They enter these "data" in the duplicate notebooks and use them to write up their reports. Since they understand the experiment well enough to create appropriate "data" and have nothing odd to explain, their reports are always clear and concise, and they receive A's. In fact, the professor has used Able and Baker's success as an example to the class of good scientific writing and of the rewards of thorough preparation before the lab session.

Questions for Discussion

1. Do you have any objections to Able and Baker's approach to taking undergraduate science laboratory classes? If so, what are they?

2. Is Able and Baker's approach ethically sound? Why or why not?

■ Case 2

Charlie Chase is also a student in Able and Baker's biology class. Chase showed up for the first lab session without even having bought the lab manual, much less read it. The first lab was carried out individually, and Chase, borrowing the associate instructor's manual as well as paper on which to record his data, was the last to finish. Needless to say, the professor was less than pleased with Chase's participation in lab that day and let his feelings be known.

Later that evening, Chase plotted the data for absorbance of various dilutions of a standard dye solution and used the resultant standard curve to determine the concentration of dye in the unknown sample. He was pleased that the plot of absorbance versus concentration yielded a line as expected and that the concentration he got for the unknown was very close to that determined by other students taking the course who lived on his floor.

Unfortunately, Bob Baker pointed out that Chase's line didn't go through zero and must therefore be wrong. What to do? Charlie wondered and worried. Suddenly he noticed a scribble he had made on the side of his data sheet, "Dropped cuvette and got new one." "Of course," he thought, "I broke one of the cuvettes I was using right after I'd finally gotten the blank all figured out and I never rechecked the blank with the new cuvette. Well, that's easy to fix."

In writing up the report, he reported his original data and then subtracted what he called a "blank factor" from each absorbance reading to generate the corrected data with which he prepared the standard curve. He explained why the "blank factor" was needed, and now had a line that went through zero.

In grading the report, the professor agreed with Chase's evaluation and treatment of the data but noted that preparation before the lab would have eliminated the hurrying that had caused both the breakage and the omission of a new blank reading. He therefore gave Chase a B.

Questions for Discussion

1. Do you have any objections to Chase's approach to taking undergraduate science laboratories? If so, what are they?
2. Is Chase's approach ethically sound? Why or why not?
3. Do you have any objections to the grade Chase received on his laboratory report? If so, what are they?

■ Case 3

Another student in the biology lab course with Able, Baker, and Chase is Donna Donovan. She came to the lab prepared and did reasonably well during class: no breakage, no crises. When she sat down to write up the results of the dye dilution experiment, she discovered that she could draw a nice, straight line through zero and almost all of the data points that should have yielded the standard curve. However, two of the points were far off the line, one very much above it and the other very much below it. How was she going to explain this in her report? Like Able and Baker, she was also desperate to do well in the course, but unlike them she didn't have time for a lot of fancy planning. After some thought, Donovan decided that the best thing would be to adjust the data so that the troublesome points would now fall very close to the line derived from the other points. That would make the report much easier to write and she would then have time to study for tomorrow's calculus quiz.

Questions for Discussion

1. How would you evaluate Donovan's actions relative to those of Able and Baker in Case 1 and Chase in Case 2?

2. Is this case more like Case 1 or Case 2?

■ Case 4

Professor Edward East is an organic chemist in the chemistry department of an agricultural college. He has done a great deal of research on the synthesis of new herbicides and is quite conversant with the literature in the field. He has recently developed a theory that he thinks will facilitate the ability to predict which derivatives of known herbicides will be effective and which will not be effective. In order to test his theory, he has synthesized 100 compounds and evaluated the effect of each on test plants. All of the results appear to support his theory.

While writing up these data for publication, Professor East discovers that he failed to include results for one of the compounds in his summary table. Since all of the raw data are in his notebooks, he quickly retrieves them, only to discover that the results for compound 84 are not as expected. His theory, supported by the results for all 99 other compounds, predicts that compound 84 should be ineffective. However, the plants treated with it did not have the expected normal growth but instead were somewhat sickly.

East is puzzled. He would like to test compound 84 again, but it is all gone and would have to be resynthesized. So, figuring that the plants probably were a bit sickly for reasons unrelated to the treatment, he decides to enter compound 84 in the manuscript's table as having no effect and publishes the results supporting his new theory.

Questions for Discussion

1. What are the similarities and differences between this case and the previous cases, particularly Case 3 (Donovan)? Be prepared to discuss their significance to your evaluation of the ethical issues involved.

2. Is it significant that East is a professional scientist doing original research rather than a student doing a class laboratory experiment? Why or why not?

■ Case 5

Continuing our discussion of Dr. Edward East and his testing of potential herbicides from Case 4, assume that this time, instead of altering his data for compound 84, Dr. East decides to publish his theory with only 99 rather than 100 supporting compounds. After all, 99 is a pretty impressive number and he figures he ought to get credit for all the work he has done. He reasons that if he is right, compound 84 will test as expected when he can find the time to synthesize more. On the other hand, if compound 84 really is an exception to his theory, he can pursue that angle and publish another paper refining the theory and adding to his publication list.

Questions for Discussion

1. Are East's actions in this case more like those of Able and Baker in Case 1 or Chase in Case 2?

2. Is data omission a less serious breach of scientific ethics than data alteration?

■ Case 6

The final experiment of an introductory biology lab course presents the students with the opportunity to do a "real" experiment, one in which no one may know the "correct" results. The students test a variety of compounds supplied by the instructor for mutagenicity in specially engineered strains of bacteria, a procedure commonly referred to as the Ames test. The "real" part of the experiment is that students are encouraged to bring things from home and test them for mutagenicity as well.

Jill Jamison decides to test black hair dye, saccharine, mushrooms, red wine, and the black crust of a flame-broiled steak. Her results indicate that all are mutagenic, at different potencies, except for saccharine. That is not what she expected after hearing all the controversy about artificial sweeteners in the media. She believes something must be wrong with her results and is not sure what to do. Thinking that the instructor will view her saccharine result as wrong, she decides to omit it from her report and only show data for the other four substances. After all, the testing of additional compounds was optional and the instructor doesn't know what she tested.

Questions for Discussion

1. Is Jamison's omission of data the best course of action?

2. Is Jamison's action unethical? Be prepared to explain your reasoning, based on similarities with previous cases, particularly Case 5.

Case 7

Dr. Karen Kenworth has been involved in the synthesis and testing of new antibiotics at Kaylabs for 20 years. Her current project involves laboratory testing of a new derivative of the company's effective and successful antibiotic Kayfab as a prelude to clinical trials involving human subjects. Everyone in the field knows which derivatives should work and which should not, based on a theory published by Kenworth 15 years ago that has come to be known in the trade as Kenworth's Law. In those 15 years, no data have been published to contradict the law.

Unfortunately, derivative 672 is giving Dr. Kenworth some trouble. Most of the results indicate that 672 is slightly more effective than Kayfab, but some show it to be much less effective. When averaged together, the data indicate that derivative 672 is a bit less effective than Kayfab. Based on the majority of the data and her time-tested law, Dr. Kenworth can't believe these odd results are valid. Since repeating the tests would take at least a year, Dr. Kenworth decides to omit the "bad points" and publish the remainder of the test data to support her conclusion that derivative 672 is slightly more effective than Kayfab.

Questions for Discussion

1. Are Dr. Kenworth's actions reasonable? Why or why not?
2. Are Dr. Kenworth's actions ethical? Be prepared to explain your reasoning based on the previous cases.
3. Recall that the next step for testing derivative 672 is human clinical trials. Is this fact significant to your evaluation? Explain.

Case 8

Gary Grant is a graduate student in the laboratory of Professor Hans Hansen. Grant's project is to determine the factors that are important for the expression of a foreign gene in plants. He is using a bacterial gene, *lac*, introduced into tobacco plants and is assaying the amount of RNA transcribed from the foreign *lac* gene. Among 20 plants he has regenerated with one arrangement of the bacterial gene, two have no detectable *lac* RNA, 15 have low levels, and three have high levels. On the pages in his notebook where these data are recorded, he has written his preliminary conclusion: "Only two of the plants fail to express *lac*; all the rest show significant expression."

As part of this project, the Hansen lab is collaborating with Dr. Ian Ingalls, a new professor in the department who is considered by all to be very bright and destined for greatness. The Ingalls lab has been assaying for the presence of the foreign *lac* protein in the transgenic tobacco plants using antibodies.

When Grant, Hansen, and Ingalls get together to compare data in preparation for publication, Ingalls says that writing the paper will be easy because only three of the tested plants express *lac*. Grant says that while that may be true for the protein, RNA analysis indicates that 18 of the plants have *lac* RNA and so they cannot say that the gene is expressed in only a few plants. Grant suggests that the protein assay is not as sensitive and should probably be rechecked. Ingalls becomes very upset, saying that he did those experiments himself, knows they were right, and doesn't need a graduate student to tell him how to do science. He then demands to see Grant's data, which Grant readily produces. Seeing the lower levels for the 15 plants in question, Ingalls addresses himself to Hansen, saying that these data support what he's just said, that expression is only observed in three plants, and Grant just doesn't know how to interpret data. Grant objects, but Hansen agrees that the low levels detected may be very close to background levels. As the discussion heats up, Hansen eventually excuses Grant from the room, as well as from all subsequent work on the paper.

When published, the paper lists Grant as a coauthor, reports the RNA data in a table as "-" for 17 plants and "+" for three, and concludes that the foreign *lac* gene is expressed in only three plants.

Questions for Discussion

1. Compare the details of this case with those of Case 3 (Donovan) and Case 4 (East). To which case is this one most similar? Are there significant differences?

2. In trying to determine whether this is an example of data alteration or of differing interpretations of the data, is there any additional information that would be useful?

7 The Ethics of Genetic Screening and Testing

THE HUMAN GENOME Project is an ambitious attempt to locate precisely and eventually sequence all of the genes in the human genome. A first stage, which is well under way, is to generate a restriction site map of all 24[1] of the human chromosomes, as well as to identify a large number of restriction enzyme cleavage sites that show variability—that is, sites which are present in some individuals and absent in others. Such variable sites result in restriction fragment length polymorphisms, or RFLPs (usually pronounced "rif-lips"), when a blot generated from the cleaved and size-fractionated DNA of different individuals is probed with a labeled DNA sequence from the region of the polymorphism.

RFLPs are transmitted exactly like alleles of a gene: that is, there will be one copy of the chromosomal interval on each of the two homologous chromosomes. During meiosis, they will separate and one copy will be incorporated into the sperm or egg cell. Thus, if a parent is heterozygous for a RFLP, the RFLP patterns of the child can be analyzed to determine which of the parental chromosomes was inherited. The same determination can also be made for a fetus by amniocentesis.

RFLP analysis is important in two situations. The first is in identifying an individual, as in paternity testing or for use as evidence in a criminal investigation. If enough RFLPs are analyzed, they will indicate a unique genetic "fingerprint" (except in identical twins). The second use of RFLPs is in screening for specific genetic disorders. Since recombination occurs along the chromosomes, the diagnostic use of RFLPs requires that the polymorphic site be closely linked to the defective gene. Research has resulted in the identification of RFLPs closely associated with the genes responsible for a number of genetic diseases.

Genetic testing has been available in a limited form for nearly 30 years. Gross chromosomal abnormalities, such as Down syndrome (caused by the existence of an extra copy of chromosome 21), can be detected by culturing fetal cells obtained from the amniotic fluid and staining the chromosomes. A few disorders can be detected by the presence of abnormal levels of a substance in the blood or other tissue. For instance, phenylketonuria (PKU) can be detected by performing a blood test on newborns, and its attendant mental retardation can be prevented by feeding affected children a restricted diet while their brains develop. PKU testing of newborns is mandatory in the United States.

RFLP analysis promises to expand enormously the number of genetic loci that can be tested. Already, genes have been located for a number of genetic diseases, such as Huntington disease, muscular dystrophy, cystic fibrosis, retinoblastoma, and familial colon cancer. The number is ever-increasing.

■ Case 1: Genetic Testing for Beta-thalassemia

Beta-thalassemia major (originally known as Cooley's anemia) is a recessive disease caused by a defective beta-globin gene that affects the ability of the red blood cells to transport oxygen. The disease does not display any effect on newborns but becomes increasingly evident with age. It is 100 percent fatal. Those who are afflicted are homozygotes; they require frequent blood transfusions and have severely reduced life spans. (They generally die by the age of 18 or 20 from a variety of complications, including autoimmune deficiency disease and liver failure). Heterozygous carriers are almost always normal, having, at most, a mild anemia. Unlike diseases such as Tay-Sachs disease and sickle-cell anemia, which are specific to a particular ethnic or racial group, beta-thalassemia affects a wide range of groups.

For the purposes of this case, assume that RFLP identification of the beta-thalassemia allele is 100 percent accurate. That is, the site of the DNA polymorphism that generates the RFLP is in the beta-globin gene itself and thus never separates by recombination. Healthy individuals can be tested and told whether they carry the defective allele. (Note: Unaffected siblings of a thalassemic [homozygous] individual have a 67 percent chance of carrying the defective allele.)

Consider the case of Alex and Maria Alwright. They have three children. Their two daughters, Sara and Jane, aged seven and four, have always been healthy, aside from normal childhood diseases. However, their two-month-old son, Chris, is often sick and is not gaining weight at a normal rate for infants his age. After Alex and Maria take Chris to a number of pediatricians who cannot ascertain what is wrong, a pediatric specialist, Dr. Wu, diagnoses him as having beta-thalassemia major.

Alex and Maria are concerned that their daughters may also have the disease. Dr. Wu tells them not to worry, that Sara and Jane are probably, at worst, heterozygous carriers for the disease. "Remember," he tells them, "both of you are healthy, and we now know you're carriers." He suggests that they might want to have the girls tested in order to know for certain whether or not they carry the defective allele.

Alex and Maria cannot agree about whether to have their daughters tested. Maria is against it, arguing that no one should have to bear the burden of being a known carrier of such a terrible disease. "Besides," she tells Alex, "we can't afford it, especially after all the medical expenses we owe for Chris's diagnosis." Alex disagrees; he thinks it's a good idea for the girls to be tested for beta-thalassemia. He also plans to recommend it to his brother, who is getting married next month. He agrees that they cannot afford to have the girls tested now but thinks it is something they should consider in the future, especially before Sara and Jane have any children of their own.

Questions for Discussion

1. Should Alex and Maria have their daughters tested for beta-thalassemia if they can find the money to do so? What ethical issues does their decision involve? Are the issues different for Alex's brother?

2. Should voluntary testing and counseling be available to all potential carriers of beta-thalassemia?

3. Many people do not realize that the healthy siblings of a child afflicted with a genetic disease may transmit that disease to their own offspring—particularly if they marry relatives (even distant ones). Because of this possibility, should individuals at risk for transmitting severe genetic diseases be required to undergo genetic testing and counseling?

4. How is the beta-thalassemia case different from that of PKU mentioned in the introduction to this chapter?

5. What are the economic implications of testing and counseling for a disease such as beta-thalassemia (e.g., who should pay and how much)? Should these implications bear on the appropriateness of testing?

6. Who should have access to the results of testing for beta-thalassemia? Does your answer depend on the results of the test? Why or why not?

7. Instead of the facts as given above, imagine that the nearest diagnostic RFLP for the beta-thalassemia gene is four map units away from the gene. This means that about 4 percent of the time the RFLP analysis will make the wrong prediction: a carrier will be misidentified as a noncarrier or vice versa. Similarly, if used in prenatal diagnosis, about 2 percent of the heterozygous fetuses will be incorrectly identified as affected homozygotes and about 2 percent of the affected homozygotes will be misidentified as healthy carriers. Does this change in facts warrant any different responses than those you gave above?

8. Sometimes research is wrong. Researchers recently thought they had identified the gene responsible for manic depression using statistical evidence derived from studies with family members. The collection of additional data, however, invalidated their original conclusion. What level (percentage) of certainty should a genetic counselor have before informing patients that they are carriers of a genetic disease?

■ Case 2

Suzanne is pregnant with her second child. She decides to have the fetus tested for cystic fibrosis, since both she and her husband are carriers of the disease. (If two heterozygotes have a child, it will have a 25 percent chance of developing the disease.) The couple's first-born child, Charisse, has been tested, and has also been diagnosed as a carrier of cystic fibrosis.

The tests indicate that the fetus, if carried to term, will be born with cystic fibrosis. Suzanne decides not to abort the fetus. Her insurance company now refuses to pay for any medical treatment for either pre- or postnatal care for her and her baby.

Questions for Discussion

1. Was the insurance company justified in withholding payment for Suzanne's pre- and postnatal care?

2. What if the genetic defect was for a disease like Huntington disease, which may not reveal symptoms for more than forty years? What if the defect was for a disease for which some type of treatment was available (e.g., retinoblastoma)?

3. Suppose that the tests indicated that Suzanne's fetus would be born normal but the child is born with cystic fibrosis. Should Suzanne and her husband have a legal claim against the doctor for "wrongful birth" if Suzanne could legitimately claim that she would have had an abortion if the fetus had been properly diagnosed?

4. Should Suzanne's decision not to have an abortion be considered child abuse, since she knows her fetus will be born impaired and will have to suffer the pain of numerous medical procedures to prolong its life?

■ Case 3

Consider the following hypothetical case:

By studying many families, researchers have determined that "hypersusceptibility" to the risk of genetic abnormalities developing as a result of lead exposure is inherited as a dominant Mendelian trait. (By "hypersusceptibility," we mean that affected individuals have increased susceptibility to conditions that have minimal effects on "normal" individuals.) Now researchers have developed RFLP screening to determine which persons have the rare allele responsible for this condition. Assume, for this hypothetial case, that the accuracy of the RFLP diagnosis for hypersusceptibility to lead mutagenicity is 100 percent.

The Power Force Company manufactures, among other products, batteries that contain lead. Power Force wants to implement a mandatory job-screening program for both current and prospective employees to determine whether they have the trait for hypersusceptibility to lead.

Questions for Discussion

1. Do you see any ethical problems with the Power Force Company's proposed lead hypersusceptibility screening program? Why or why not?

2. Does it matter whether the company is seeking to screen current, as opposed to prospective, employees? Explain.

3. Of what relevance, if any, to your analysis is the company's intended use of the information gained by its proposed screening? For example, do you see any ethical problems with any of the following actions? If so, what are they?

   a. Excluding current hypersusceptible employees from the workplace entirely (including firing current employees who have this trait);
   b. Refusing to hire prospective employees with the trait;
   c. Giving hypersusceptible persons the option of working in an environment where lead is present, or, alternatively, being relocated into a different area of the company;
   d. Testing employees, but not giving them the results of the tests;
   e. Using the information to determine the employees' eligibility for the company's health insurance program.

4. Should the ethical acceptability of workplace screening require that the screening be for a genetic susceptibility that is related to exposure to hazards present in the workplace (as opposed to testing for other genetic problems that are not directly linked to the job)? Why or why not?

5. What if the results of screening reveal that only a certain group of persons bear the trait for hypersusceptibility to lead? Would you change your answer if the only affected group were (a) persons of black African ancestry; or (b) persons of German ancestry? If so, how?

6. Consider the following possibility: What if further research in the field of lead mutagenicity revealed that individuals with the trait are only hypersusceptible before birth? That is, a woman heterozygous for the hypersusceptibility allele is now, herself, no more susceptible than any other member of the population, but if a fetus she carries has inherited the allele from her (for which there is a 50 percent probability), it will be hypersusceptible to mutation if exposed to lead.

Would the Power Force Company be justified in requiring that any woman of childbearing age who had the hypersusceptibility allele and wanted to work in an environment containing lead first show proof of infertility or sterilization? Most people are unlikely to know their genetic makeup. Should all women of childbearing age who want to work in lead-containing environments be tested?

Note that the fetus could have inherited the hypersusceptibility allele from its father as well. Could the Power Force Company justifiably insist that *any* woman of childbearing age who wanted to work in a lead-containing environment first show proof of infertility or sterilization?

■ Case 4

Jay is a 20-year-oid college junior. His maternal grandfather has recently been diagnosed as having Huntington disease (HD). HD is a late-onset genetic disease whose symptoms—severe mental and physical deterioration—do not appear until adulthood (usually after age 40). There is no effective treatment or cure for HD. Genetic researchers have identified and cloned the gene that is responsible for HD so that diagnosis based on DNA analysis is 100 percent accurate. Jay's mother has decided to be tested and wants Jay to be tested as well.

Questions for Discussion

1. Should Jay be tested for Huntington disease? If so, should he be informed of the results? Why or why not? If you were Jay, would you want to be tested? Would you want to know the results? Are your reasons the same or different from those you gave with respect to Jay? What if Jay were ten years old?

2. Would your answers to Question 1 be different if the proposed genetic testing was for a disease such as manic depression, whose symptoms reveal themselves at a much earlier age than HD?

3. Would your answers to Question 1 be different if HD were treatable?

4. Would your answers to Question 1 be different if pre-symptomatic testing and diagnosis could lead to a breakthrough in effective treatment or cure for HD?

5. Should family genetic histories be incorporated into the standard medical histories required for various purposes (e.g., school or college enrollment, employment, insurance)?

6. The gene responsible for HD is one of the so-called trinucleotide repeat genes. These genes include long stretches of three DNA nucleotides repeated over and over. It appears that these repeated arrays can expand as the gene is passed on in a family. When the array becomes too large, the gene appears to become defective and the disease is manifest.

Researchers have speculated that the length of the repeated array may correlate with the severity of the disease. In the case of HD, it is thought that the longer the array, the earlier the onset will be.

Assume for the moment that this is true, and that human geneticists can predict not only whether a person will get HD but also the approximate age at which the symptoms will begin.

Should physicians be willing to make such a determination if their patients request it? Should it be a standard part of the HD testing procedure? If such information were to become available to people, what are some of the implications for physicians and their responsibilities toward their patients (e.g., counseling and follow-up care)?

Case Notes

The following is an excerpt from a report of the U.S. President's Commission for the Study of Ethical Problems in Medicine and Biomedical and Behavioral Research, *Screening and Counseling for Genetic Conditions* (1983: 5–8). Consider the Commission's conclusions as they relate to the cases in this chapter.

The Commission's basic conclusion is that programs to provide genetic education, screening, and counseling provide valuable services when they are established with concrete goals and specific procedural guidelines founded on sound ethical and legal principles. The major conclusions fall into five categories.

Confidentiality

(1) Genetic information should not be given to unrelated third parties, such as insurers or employers, without the explicit and informed consent of the person screened or a surrogate for that person.

(2) Private and governmental agencies that use data banks for genetics-related information should require that stored information be coded whenever that is compatible with the purpose of the data bank.

(3) The requirements of confidentiality can be overridden and genetic information released to relatives (or their physicians) if and only if the following four conditions are met: (a) reasonable efforts to elicit voluntary consent to disclosure have failed; (b) there is a high probability both that harm will occur if the information is withheld and that the disclosed information will actually be used to avert harm; (c) the harm that identifiable individuals would suffer if the information is not disclosed would be serious; and (d) appropriate precautions are taken to ensure that only the genetic information needed for diagnosis and/or treatment of the disease in question is disclosed.

* When it is known in advance that the results of a proposed screening program could be uniquely helpful in preventing serious harm to the biological relatives of individuals screened, it may be justifiable to make access to that program conditional upon prior agreement to disclose the results of the screening.

(4) Law reform bodies, working closely with professionals in medical genetics and organizations interested in adoption policies, should urge changes in adoption laws so that information about serious genetic risks can be conveyed to adoptees or their biological families. Genetic counselors should me-

diate the process by which adoptive records are unsealed and newly discovered health risks are communicated to affected parties.

Autonomy

(5) Mandatory genetic screening programs are only justified when voluntary testing proves inadequate to prevent serious harm to the defenseless, such as children, that could be avoided were screening performed. The goals of "a healthy gene pool" or a reduction in health costs cannot justify compulsory genetic screening.

(6) Genetic screening and counseling are medical procedures that may be chosen by an individual who desires information as an aid in making personal medical and reproductive choices.

* Professionals should generally promote and protect patient choices to undergo genetic screening and counseling, although the use of amniocentesis for sex selection should be discouraged.
* The value of the information provided by genetic screening and counseling would be diminished if available reproductive choices were to be restricted. (This is a factual conclusion that is not intended to involve the commission in the national debate over abortion).

Knowledge

(7) Decisions regarding the release of incidental findings (such as nonpaternity) or sensitive findings (such as diagnosis of an XY-female) should begin with a presumption in favor of disclosure, while still protecting a client's other interests, as determined on an individual basis. In the case of nonpaternity, accurate information about the risk of the mother and putative father bearing an affected child should be provided even when full disclosure is not made.

(8) Efforts to develop genetics curricula for elementary, secondary, and college settings and to work with educators to incorporate appropriate materials into the classroom are commendable and should be furthered. The knowledge imparted is not only important in itself but also promotes values of personal autonomy and informed public participation.

(9) Organizations such as the Association of American Medical Colleges, the American Medical Association, and the American Nursing Association should encourage the upgrading of genetics curricula for professional students. Professional educators, working with specialty societies and program planners, should identify effective methods to educate professionals about new screening tests. Programs to train health professionals, pastoral counselors, and others in the technical, social, and ethical aspects of genetic screening deserve support.

Well-being

(10) A genetic history and, when appropriate, genetic screening should be required of men donating sperm for artificial insemination; professional medical associations should take the lead in identifying what genetic infor-

mation should be obtained and in establishing criteria for excluding a potential donor.

* Records of sperm donors are necessary, but should be maintained in a way that preserves confidentiality to the greatest extent possible.
* Women undergoing artificial insemination should be given genetic information about the donor as part of the informed consent process.

(11) Screening programs should not be undertaken unless the results that are produced routinely can be relied upon.

* Screening programs should not be implemented until the test has first demonstrated its value in well-conducted, large-scale pilot studies.
* Government agencies involved in introducing new screening projects should require appropriate pilot studies as a prerequisite to approval of the product or to the funding of services.
* Government regulators, funding organizations, private industry, and medical researchers should meet to discuss their respective roles in ensuring that a prospective test is studied adequately before genetic screening programs are introduced.

(12) A full range of prescreening and follow-up services for the population to be screened should be available before a program is introduced.

* Community leaders and local organizations should play an integral part in planning community-based screening programs.
* State government should consider establishing a review group with professional and public members to oversee genetic services.
* New screening programs should include an evaluation component.

Equity

(13) Access to screening may take account of the incidence of genetic disease in various racial or ethnic groups within the population without violating principles of equity, justice, and fairness.

(14) When a genetic screening test has moved from a research to a service delivery setting, a process should exist for reviewing implicit or explicit policies that limit access to the genetic service; the review should be responsive to the full range of relevant considerations, to changes in relevant facts over time, and to the needs of any groups excluded.

* The time has come for such a review of the common medical practice of limiting amniocentesis for "advanced maternal age" to women 35 years or older.

(15) Determination of such issues as which groups are at high enough risk for screening or at what point the predictive value of a test is sufficiently high requires ethical as well as technical analyses.

(16) Cost-benefit analysis can make a useful contribution to allocational decision making, provided that the significant limitations of the method are clearly understood; it does not provide a means of avoiding difficult ethical judgments.

Note

1. The 24 human chromosomes are composed of 22 autosomes plus the two sex chromosomes, X and Y.

Reference

U.S. President's Commission for the Study of Ethical Problems in Medicine and Biomedical and Behavioral Research. *Screening and Counseling for Genetic Conditions*. Washington, DC: Government Printing Office, 1983.

8 Ethics and Eugenics

THE THREE CASES in this chapter raise ethical issues about eugenics primarily in connection with pedigree analysis. Some of the discussion questions are designed to point out parallels between the eugenics movement that was active in the early decades of this century and the choices we as a society will be called upon to make in the near future concerning the emerging technologies of genetic screening, prenatal and preimplantation testing, and gene therapy. There are historical lessons to learn that may guide us in our future decisions.

Eugenics is defined as the use of selective breeding according to the principles of genetics to remove bad traits (negative eugenics) and to spread good ones (positive eugenics) in the population. The eugenics movement was enormously influential in the first part of the twentieth century as an attempt to improve the human race scientifically. Although there was disagreement among researchers concerning the heritability of certain traits, especially that of intelligence, the eugenics movement was sufficiently influential to have far-reaching consequences.

A dominant view about eugenics is reflected in this statement by Supreme Court Justice Oliver Wendell Holmes affirming the forced sterilization of Carrie Buck, a young, feebleminded woman committed to a state institution:

> We have seen more than once that the public welfare may call upon the best citizens for their lives. It would be strange if it could not call upon those who already sap the strength of the state for these lesser sacrifices, often not felt to be such by those concerned, in order to prevent our being swamped with incompetence. It is better for all the world, if instead of waiting to execute degenerate offspring for crime, or to let them starve for their imbecility, society can prevent those who are manifestly unfit from continuing their kind. The principle that sustains compulsory vaccination is broad enough to cover cutting the Fallopian tubes. Three generations of imbeciles are enough.[1]

The eugenics movement led to the passage of laws resulting in the involuntary institutionalization or sterilization of thousands of Americans to prevent them from passing on what were presumed to be genes for feeblemindedness or criminal behavior. In 1949, 27 states had some form of eugenics laws. From 1907, when the first eugenics sterilization act was passed (in Indiana), through 1948, 49,207 eugenic sterilizations were performed in this country. Forty-five percent of the sterilizations were for insanity and 51 percent were for feeblemindedness. The remainder were for epilepsy, birth defects, or sexual immorality (Woodside 1950: 20–22). Nineteen thousand of the forced sterilizations were performed in California, one of the most ethnically diverse states in the country.

In addition, U.S. immigration laws were changed to deny entry to persons with "undesirable genes." Furthermore, eugenics influenced the incorporation of a quota system into immigration policy, which for many years determined the number of persons from different geographical areas who would be allowed to immigrate to the United States.

The accepted scientific principle on which the eugenics movement was based was that complex social behaviors (e.g., criminality) or talents (e.g., musical genius) could be inherited in a simple Mendelian fashion. Support for this concept came from pedigree analysis—determining the phenotypes of family members and arguing from similarity. But the analyses carried out by the proponents of eugenics were flawed by subjective determinations of phenotypes, the assumption that these traits are monogenic, and failure to consider the role of environment in shaping phenotypes.

Pedigree analysis is a perfectly valid method for determining inheritance of certain traits in humans. In plants and animals, we can measure objectively the phenotypes of large numbers of individuals from genetically defined strains in a carefully controlled environment. In humans, however, we are limited to the study of existing individuals.

Many traits, such as hemophilia, blood groups, and polydactyly (extra fingers) are determined by single genes. Their mechanism of inheritance has been determined accurately by pedigree analysis. This works because these are simple traits whose expression is not influenced by the individuals' environments.

It is a truism that people inherit not only their genes but also their environment. Thus, for example, a child will learn to read only if it is raised in a literate society. Eugenicists largely discounted the effects of environment on a person's achievements. The conclusions of eugenics were based on a combination of poor research techniques, wishful thinking, and prejudice. Clearly, there are inherent differences in intelligence, personality, and talent, even in families where the environment is similar for all the siblings. We know nothing, for example, about the genetic factors which can cause a difference in intelligence quotient (IQ) of twenty points between two siblings. There are many single genes which can cause large deviations from the average, including those genes responsible for phenylketoneuria and Down syndrome.

It is still undetermined how much variation within a population arises from genetic differences and how much from environmental influences. One danger posed by the use of science to guide public policy is the popular perception that the results of scientific inquiry represent "the truth." In fact, science is the search for truth, and sometimes the accepted wisdom of one generation will be discarded or dramatically modified in the next.

The following cases are based on lectures and textbooks written in the 1930s and 1940s. They are presented for practice in pedigree analysis, as a stimulus to critical thinking about the nature of proofs, and to encourage thought about the appropriate role of science in society.

■ Case 1: The Naval Heroism Gene

Consider the pedigree of a family of military heroes (Pedigree 1).[2] Henry Goddard and other eugenicists argued that this pedigree demonstrated a genetic basis for the disproportionately large number of military heroes in the Porter family. (A black box designates that the person was a naval officer; an X in a box indicates that the person was in the army.)[3]

Questions for Discussion

1. Assume this phenotype is indeed the expression of a single gene. What is the most likely mode of transmission of the naval heroism gene?

2. What are some of the factors that might have affected the choice of these men to go into the military and the likelihood that they would become officers? What do you think is the relative importance of each of these factors? Which of each of these factors is likely to be genetic, environmental, or a combination of the two?

3. What degree of certainty should researchers have about the inheritance of a particular trait (that is, the conclusion that the trait was genetically determined) before publishing their findings? What considerations are relevant to your answer?

4. When this pedigree was presented, the researchers argued that for the bearers of such desirable traits "it is important that the consort's stock shall not carry undesirable genetic factors that shall cancel the desirable traits of the propositus, but on the contrary shall carry factors of a sort that make desirable their reproduction in full or even increased degree. Proper mate selection is the key to improvement of the race."[4]

Suppose that a dating service advertises itself as requiring genetic screening of all clients so as to generate a scientific match of persons who are genetically compatible and thus able to produce the most intelligent and talented children possible. What are the implications of such a requirement? Is this a form of eugenics? Is this a socially desirable use of genetic screening?

Suppose that a sperm bank has a policy that prohibits a woman seeking artifical insemination access to its stores of sperm from men considered to be geniuses if the woman fails to meet certain criteria for intelligence. Is this a form of eugenics? What are the implications of such policies for genetic research and for society?

5. If our investigations of the human genome lead to the identification of some alleles that tend to be associated with a desirable trait, such as musical

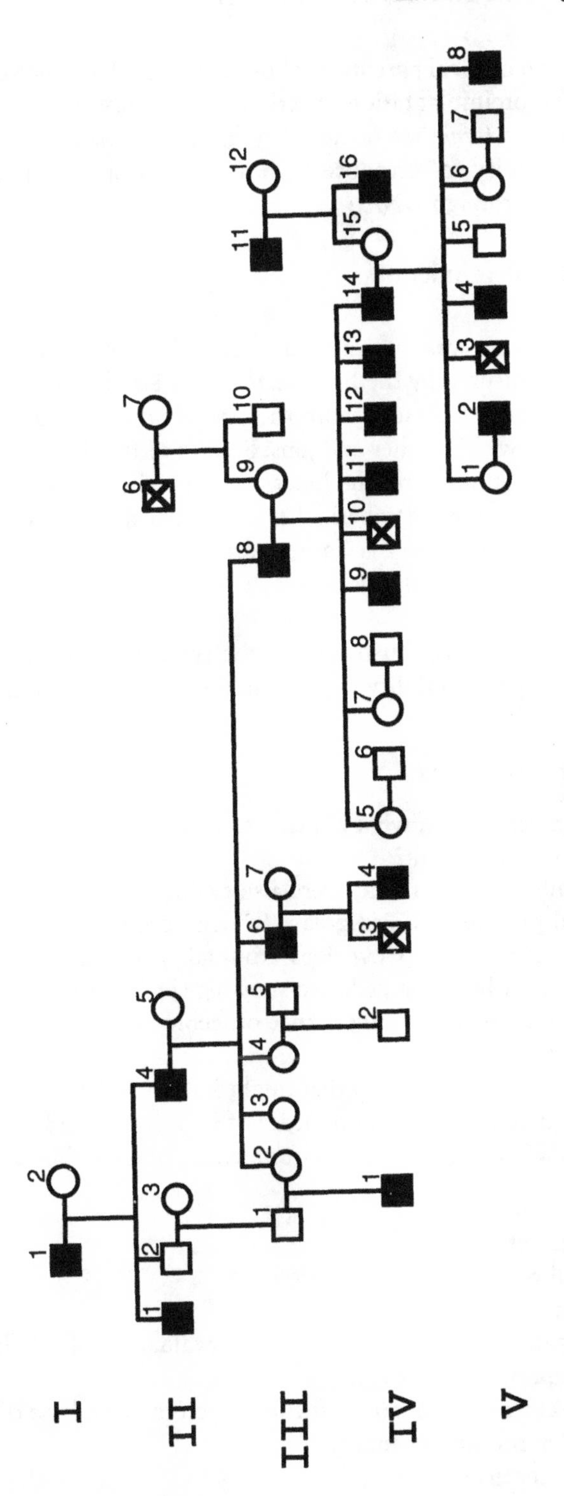

Pedigree 1. Porter pedigree (naval heroism).

talent, should two musical parents be allowed to use inheritance of these alleles as a criterion for preimplantation selection of an embryo?

Should parents (whether or not they have expressed desirable traits) be given the opportunity to introduce such alleles into a preimplantation-stage embryo (i.e., from foreign DNA)?

■ Case 2: Feeblemindedness

Feeblemindedness was a term used to describe anyone with a mental age of 12 or less as measured by the Binet test[5] (IQ of 80 or lower). The classification covered a range of persons from the severely mentally retarded to many "normal" persons with families and jobs. Studies of two large, inbred families in the Appalachian mountains, the Jukes and the Kallikaks, were used to show that feeblemindedness is an inherited trait. (A background summary of the Kallikak study is provided in the Case Notes at the end of this chapter.) These studies provided much of the data which were used to justify the eugenics laws.

Pedigrees 2 and 3 are illustrative of the types of Jukes family pedigrees used to "prove" the heritability of feeblemindedness and criminality, respectively.[6]

Questions for Discussion

1. Based on Pedigree 2, what appears to be the mechanism of inheritance of the feeblemindedness trait?

2. The members of this family were also evaluated for criminality, alcoholism, and sexual promiscuity. Pedigree 3 (which appears on page 78) presents information on criminality. How does criminality appear to be inherited in this family? Consult both the pedigree showing the inheritance of criminality and the pedigree showing the inheritance of feeblemindedness when considering your answer.

3. Following are some factors that might influence the behavior or mental capacity of the Jukes. Rank them in order of importance and discuss whether each factor is likely to be genetic, environmental, or a combination of both.

a. lack of education

b. community mores

c. mental retardation from inbreeding

d. poverty

e. data gathering bias (e.g., subjective evaluation of intelligence based on facial expression or lifestyle)

f. data gathering errors (e.g., incorrect paternity and use of hearsay evidence for determining phenotypes)

g. other (specify)

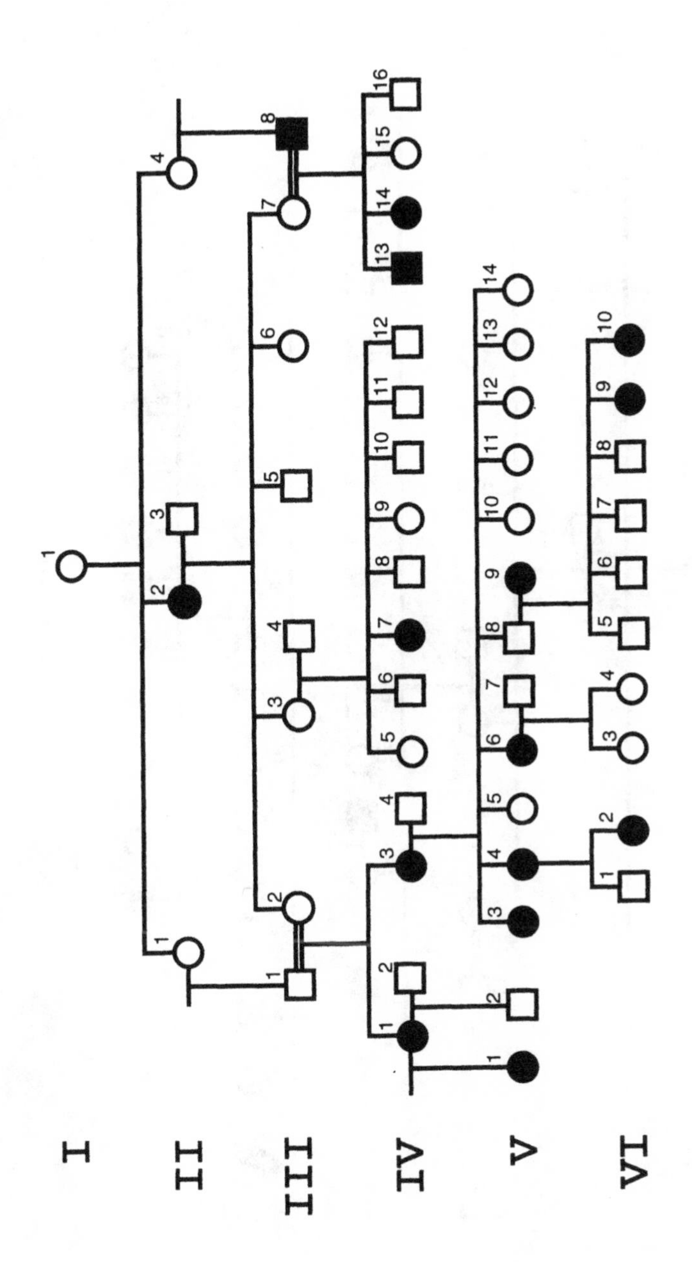

Pedigree 2. Feeblemindedness.

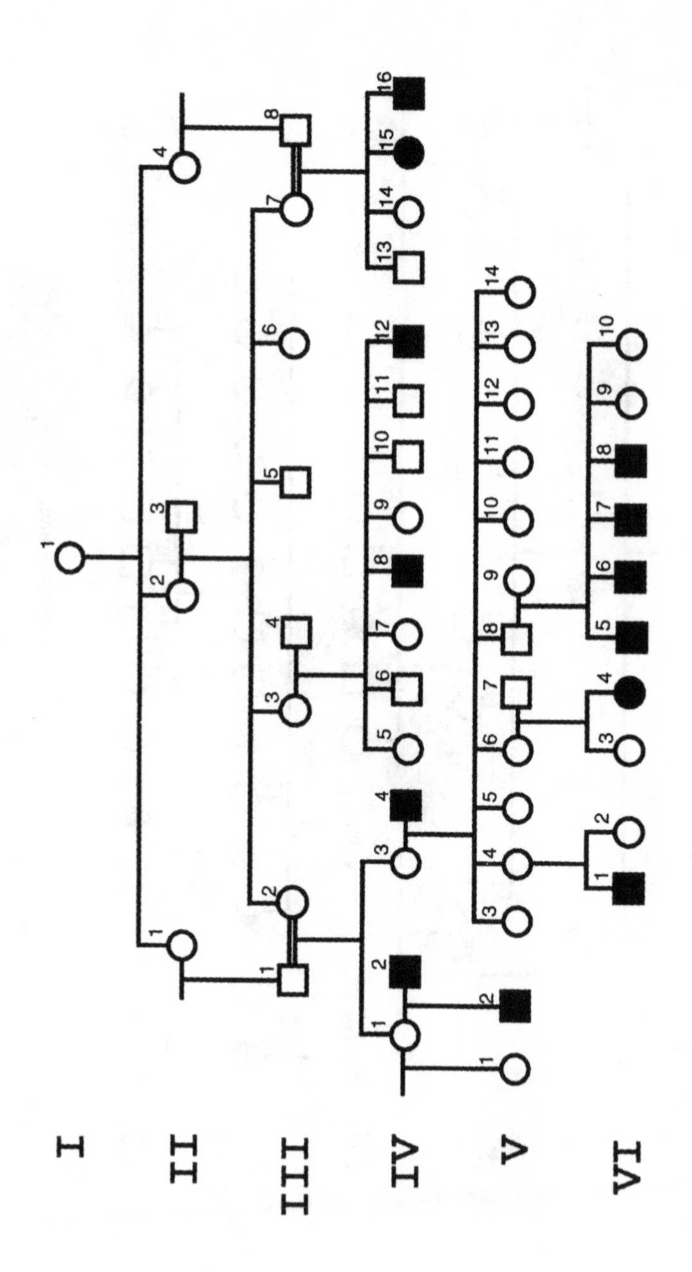

Pedigree 3. Criminality.

Given your ranking, what other factors, if any, should researchers take into consideration in doing this kind of study?

4. If the researchers studying the Jukes had been correct in their evaluation that inbreeding of this family had resulted in feeblemindedness, how should society respond? Is it appropriate for society to continue to carry the financial burden of increasing numbers of criminals, alcoholics, and mentally deficient persons if these characteristics are genetically determined?

5. What responsibility, if any, should researchers have for the ways members of society and social institutions respond to their findings? What can or should researchers do with respect to the uses to which their research findings are put?

6. Unlike criminality, we are certain that diseases such as cystic fibrosis, Huntington disease, and retinoblastoma have a genetic cause. Treatment of individuals who have genetic diseases puts a strain on our nation's limited health care budget. In addition, with each successive cloning of a gene associated with a genetic disease, it becomes easier to carry out large scale testing for individuals who carry defective alleles of these genes.

Should all individuals be required to undergo genetic testing for a number of genetic diseases as well as counseling before receiving a marriage license?

Should the reproductive choices of those carrying defective alleles be regulated? For instance, consider a law that requires couples at risk for producing offspring with a genetic disease to do everything possible to prevent the birth of an affected child or be denied insurance coverage.

■ Case 3: Ethical Considerations in Data Collection

The Kallikak study (Goddard 1912) was regarded as proof of the genetic inheritance of feeblemindedness. In 1803, Martin Kallikak had a son, Martin, Jr., by a "nameless feeble-minded girl." Martin soon left the girl and married a girl of "good" family, by whom he had seven children, all of normal intelligence. He became wealthy and left large farms to his legitimate children.

The first son (considered to be feebleminded) was the progenitor of 480 progeny in the next four generations, including 143 designated as feebleminded, 3 as criminals, 33 as "sexually immoral," and 24 as alcoholics. Martin's legitimate progeny also multiplied, resulting in a pedigree of 496 mentally normal offspring, many of them well-to-do farmers and professionals. There were only two alcoholics and no criminals among the "good" Kallikaks. (A further summary of the Kallikak study is provided in the Case Notes at the end of this chapter.)

The pedigrees and the determinations of mental capacity were made by interviewing the surviving family members and other members of the community. Some of the Kallikaks had been dead for 80 years when the study was done. Here is an excerpt from the field-worker's notebook giving her own observations.

> By dint of persistent inquiry, the family was discovered living in the back shed of a dilapidated country tenement. . . . Three children, scantily clad and with shoes that would barely hold together, stood about with drooping jaws and the unmistakable look of the feeble-minded. . . . The whole family was a living demonstration of the futility of trying to make desirable citizens from defective stock through making and enforcing compulsory education laws. . . . The father himself, though strong and vigorous, showed by his face that he had only a child's mentality. The mother in her filth and rags was also a child. . . . In this house of abject poverty, only one sure prospect was ahead, that it would produce more feeble-minded children with which to clog the wheels of human progress (Goddard 1912: 77–78).

The entire family was rated as feebleminded. Here is a second excerpt from the notebook, giving a neighbor's reminiscences of one of Martin, Jr.'s children.

> Old Moll, simple as she was, would do anything for a neighbor. She finally died—burned to death in the chimney corner. She had come in drunk and sat down there. Whether she fell over in a fit or her clothes caught fire, nobody knows. She was burned to a crisp when they found her (Goddard 1912: 79–80).

Old Moll was scored as alcoholic, epileptic, sexually immoral, and feebleminded (Goddard 1912: 22).

The chief researcher who conducted the Kallikak study, Henry Goddard, believed that the distribution of feebleminded individuals in the "bad" Kallikak lineage proved that it segregated as a single-gene recessive trait. This assumption would have required that Martin Kallikak be heterozygous for feeblemindedness. Martin's seven legitimate children should therefore have contained several heterozygotes as well, yet none of them had any feebleminded offspring.

There were five instances among the "bad" Kallikaks in which very young children were taken from their mothers and raised by "good" families. Four of these children were apparently of normal intelligence, although two of them were born to parents who were both designated as feebleminded by the study. Goddard argued that their environment was unlikely to be the relevant factor in determining their normal intelligence and that, instead, these individuals were "high-grade morons, who, to the untrained person, would seem to be . . . nearly normal" (Goddard 1912: 62).

Questions for Discussion

1. What ethical problems are there, if any, with the field-worker's observations?

2. What are the social and ethical implications of the view that traits such

as feeblemindedness are hereditary? Is your answer different from your response to the question in connection with the trait of courage/cowardice in Case 1? If so, why?

3. What degree of certainty should researchers have about the inheritance of a particular trait (i.e., the conclusion that the trait was genetically determined) before publishing their findings? What considerations are relevant to your answer?

4. What ethical problems, if any, does Goddard's conclusion about the distribution of feeblemindedness in the "bad" Kallikak lineage pose?

5. What would you conclude about the distribution of the feeblemindedness allele in the population?

6. Does Goddard's argument about "high-grade morons" raise any ethical concerns? If so, what are they?

7. Does society have the right to prevent some individuals from having children or to discourage them from doing so? Why or why not?

8. If we want to improve the human species genetically, how will we determine which traits are desirable and which are undesirable? For instance, an increased risk for developing diabetes may be undesirable, but is short stature also undesirable?

Case Notes: Summary of the Kallikak Study

In 1910, the Training School for Backward and Feeble-minded Children in Vineland, New Jersey, opened a laboratory and department of research for the study of feeblemindedness. The purpose of the research was to determine the cause of feeblemindedness. In 1912, Henry H. Goddard, the director of the research lab at the school, published *The Kallikak Family: A Study in the Heredity of Feeble-mindedness*, a study which traced the family history of Deborah Kallikak, an allegedly feebleminded young girl who had been brought to the training school when she was eight years old.

Goddard obtained information on the families of feebleminded children by sending trained workers to the children's homes. These field-workers were instructed to obtain information about the children and their families through careful and thoughtful questioning of relatives, neighbors, employers, teachers, physicians, and ministers as well as family members. The field-workers also attempted to determine if dead family members were feebleminded by using original documents (whenever possible) and any memories or records of how the person lived. Goddard claimed that "[t]hese facts are frequently sufficient to enable one to determine, with a high degree of accuracy, whether the individual was normal or otherwise" (Goddard 1912: 14).

Goddard's study of the Kallikaks contains many descriptions by Elizabeth Kite, one of the field-workers who visited these families. She describes the families as often poor and living in unsanitary conditions. Because the Kallikak family was so well known—actually, notorious in the community—the field-workers were able to trace six generations of Kallikaks with feeblemindedness and other defects. In addition, they found many normal families in the community who had the Kallikak name, and hypothesized that Deborah's family must have been the "degenerative offshoot" of an older family of "better stock" (Goddard 1912: 16).

The field-workers found that Martin Kallikak, Sr., Deborah's great-great-great grandfather, was normal and came from a family with good standing in the community. When Martin, Sr., was a young man, he met a feebleminded young woman in a tavern, and she bore Martin Kallikak, Jr., Deborah's feebleminded great-great grandfather. Martin, Sr., later married a normal woman, and all of their children, the half-brothers and sisters of Martin, Jr., were normal. In short, Goddard concluded that the family had normal generations on the good side, and mental defectives on the other, starting with the nameless girl in the tavern (Goddard 1912: 30–31).

Of Martin, Jr.'s, 480 descendants, Goddard states to have conclusive proof that 143 were feebleminded, while only 46 were found to be normal. (The rest were either unknown or of doubtful intelligence.) In addition, 36 were illegitimate, 33 were sexually immoral or were prostitutes, 24 were confirmed alcoholics, 3 were epileptics, 3 were criminal, and many more had other kinds of defects (Goddard 1912: 18–19). The study's long and detailed family histories of these people show that many had married into feebleminded families and produced a majority of feebleminded children.

In Goddard's words, "[f]eeble-mindedness is hereditary and transmitted as surely as any other character. We cannot successfully cope with these conditions until we recognize feeble-mindedness and its hereditary nature, recognize it early, and take care of it" (Goddard 1912: 117).

The conclusion of the study was that feeblemindedness is an inherited trait. Goddard offered the family histories of both sides of the Kallikak family as conclusive evidence of this judgement. According to Goddard, the enormous number of feebleminded individuals in the descendants of Martin, Jr., and the absence of feeblemindedness in the descendants of his half-brothers and -sisters proves that it is not the environment but heredity that has produced these differences (Goddard 1912: 53). (According to Mendelian theory, all of the children of two feebleminded parents should be feebleminded. But in the case of the Kallikaks, one feebleminded couple had a child who was normal and had normal children.)

Goddard further concluded that the Kallikaks' feeblemindedness was responsible for their criminal, "immoral," or alcoholic tendencies.

> We find on the good side of the family prominent people in all walks of life and nearly all of the 496 descendants owners of land or proprietors. On the bad side we find paupers, criminals, prostitutes, drunkards, and examples of

> all forms of social pest with which modern society is burdened. From this we conclude that feeble-mindedness is largely responsible for these social sores (Goddard 1912: 116).

As a solution to the problems caused by the feebleminded, Goddard suggested that these individuals be segregated and colonized. He felt that this would take the place of almshouses and prisons, actually decrease the number of mental hospitals, and reduce or eliminate the loss in property and life caused by their actions (Goddard 1912: 105–106). Goddard also proposed the idea of preventing these people from having children, either by "unsexing" them through castration or ovariectomy or by sterilizing them with a vasectomy or salpingectomy (an excision of the fallopian tube) (pp. 106–109). Goddard's conclusions were greatly influential in the eugenics laws passed in the earlier part of the twentieth century, as discussed in the introduction to this chapter.

Notes

1. *Buck v. Bell*, 274 U.S. 200, 207 (1927). Carrie Buck was the daughter of a feebleminded mother committed to the same institution and the mother of an illegitimate feebleminded child. Thus the reference to "three generations of imbeciles."

2. The pedigree is taken from Davenport 1919: 180. A version of this pedigree was also published in Davenport 1940: 32.

3. Some of these men died too young to be either officers or heroes. For the sake of argument, assume they would all have been credits to their family and uniforms.

4. Davenport 1940: 30.

5. The Binet or Binet-Simon test is a standardized test developed by Alfred Binet and Theodore Simon in the first decade of the 1900s to determine mental ability. Mental ability is measured by rating the test taker's performance in a number of tasks, ranging from knowledge of general information to ability to count, repeat phrases, recognize names, describe things, and make sense discriminations (see Davenport 1911: 65–66). The Binet test was widely used by psychologists and others in the first part of the twentieth century to classify the "mental age" of persons. One particular use of the Binet test was to identify those immigrants to the United States who were "feebleminded."

6. These pedigrees are modified versions of a figure shown in Davenport 1940: 60, originally published in Estabrook 1916: 67, chart 20.

References

Bishop, Jerry E. "Unnatural Selection." *National Forum* 73: 27–29 (1993).

Davenport, Charles B. *Medical Genetics and Eugenics*. Philadelphia: Women's Medical College of Pennsylvania, 1940.

———. *Naval Officers: Their Heredity and Development*. Washington, DC: Carnegie Institution of Washington, 1919.

———. *Heredity in Relation to Eugenics*. New York: Henry Holt, 1911.

Estabrook, Arthur H. *The Jukes in 1915*. Washington, DC: Carnegie Institution of Washington, 1916.
Goddard, Henry H. *The Kallikak Family*. New York: Macmillan, 1912. Reprinted by Arno Press, New York, 1973.
Handyside, A. H., et al. "Birth of a Normal Girl after In Vitro Fertilization and Preimplantation Diagnostic Testing for Cystic Fibrosis." *New England Journal of Medicine* 327: 905–909 (1992).
Kevles, Daniel J., and Leroy Hood, eds. *The Code of Codes: Scientific and Social Issues in the Human Genome Project*. Cambridge, MA: Harvard University Press, 1992.
Woodside, Moya. *Sterilization in North Carolina*. Chapel Hill: University of North Carolina Press, 1950.
Wexler, Nancy. "Presymptomatic Testing for Huntington's Disease: Harbinger of the New Genetics." *National Forum* 73: 22–26 (1993).

PART III

Cases in the Behavioral Sciences: Psychology

9 Ethical Issues in Animal Experimentation

THE CASES IN this chapter present ethical issues that arise in connection with scientific research involving the use of laboratory animals. It has been estimated that between 17 and 22 million animals are used each year for research, testing, and education. Since 1985, federal law has required that most experimental research involving animals receive the approval of institutional animal care and use committees (IACUCs).[1] IACUCs comprise various members of a local community, including scientists and nonscientists, professionals and nonprofessionals.[2] These committees monitor the care of experimental animals in the laboratory and review the appropriateness of research involving animals, in terms of both the number and the type of animals used in particular experiments. They also monitor the design of the experiments. (For a list of considerations used by IACUCs, see Case Note 1 at the conclusion of this chapter.)

The first case presented here involves the selective breeding of laboratory rats to develop certain traits characteristic of alcoholism. Among other issues, it raises the problem of "speciesism," that is, whether, or to what extent, humans are justified in controlling or using other species as means to accomplish ends that principally benefit humans.

The second case involves testing the recovery of function in monkeys' arms after deafferentation (depriving an animal of all sensations from a limb, including pain, by cutting the sensory nerves from the limb). In addition to speciesism, an important concern is whether "higher" species are entitled to greater protection from experiment-derived pain, suffering, and death.

As you read the following cases, think about whether, as a member of the university's IACUC, you would permit the research to go forward. Most IACUCs operate using consensus decision making. IACUCs often review cases that require them to balance the ethical costs of a research project—usually measured in terms of suffering and loss of life to animal subjects—against its proposed benefits, such as advances in knowledge and the application of such knowledge to benefit humans.

To help you in your task of evaluating the ethical acceptability of using animals for experimental purposes, we have included in the Case Notes a summary of some of the major approaches that have been developed. All of these ethical approaches prohibit animals from being treated as mere objects to be manipulated at the will of the researcher. In addition, they all mandate that researchers consider the purposes and importance of their research, less harmful alternatives, and the benefits and harms involved for both human and animal life before proceeding with any experiment using laboratory animals.

Beyond the question of the use of animals in research, another issue to consider in these cases is whether the actions of animal-rights groups to stop experiments, other than those involving formal legal channels, are justified. The question is whether the use of civil disobedience as a means of halting research with animals is appropriate. Think about whether the ethical principles at stake here are sufficiently strong to warrant acts of civil disobedience. Keep in mind that in their efforts to "liberate" animals and deter researchers from proceeding with experiments using animals, some animal-rights advocates have destroyed laboratory and personal property and harassed and threatened researchers and their families. These tactics have increased the costs and slowed the progress of research involving animals in the study of organ transplantation procedures and of debilitating and dangerous diseases, including Alzheimer's, AIDS, and heart disease (see McCabe 1986: 114). In discussing these cases, you will be asked to determine whether the behavior of the animal-rights group members went beyond the boundaries of acceptable ethical conduct.

■ Case 1: Breeding Alcoholic Rats

Read the following research proposal involving the experimental use of animals.

Part A. The Research Proposal

Title: "Stress and Alcoholism"
Investigators: John Jones and Mary Smith

INTRODUCTION

This study will utilize "alcohol-preferring" (P) and "alcohol-nonpreferring" (NP) strains of rats developed by selective breeding here at State University. These laboratory strains have been subjected to selective breeding over a 20-year period so that when raised under identical environmental conditions and given a choice between water and an alcohol solution, the P rats will choose the alcohol solution in preference to the water while the NP rats will not. Thus, for the P rats, the preference for alcohol is genetically determined. It has been shown that the P rats voluntarily drink alcohol solutions to the point of intoxication, metabolic and neuronal tolerance, and physical dependence.

Previous studies indicate that the patterns of alcohol consumption by the P rats resemble those of humans closely enough to provide a suitable model for the study of human alcoholism. We are now interested in observing the role that environmental factors play in the behavior of P and NP rats. In this experiment, we propose to study the effects of crowding-induced stress on alcohol consumption in both strains of rats.

EXPERIMENTAL DESIGN

The following description is a somewhat simplified version of the design actually proposed.

A total of 160 rats will be used, 80 each of the alcohol-preferring (P) and nonpreferring (NP) strains. All animals will be housed in separate but identical cages. One-half of the animals (40 P and 40 NP rats) will be housed in uncrowded conditions (10 rats per standard 10-rat cage); the other half will be housed in crowded conditions (20 rats per standard 10-rat cage). Thus the complete experimental design is a 2 × 2 (Strain [P vs. NP] by Stress [crowded vs. uncrowded] independent groups design, as diagrammed in figure 1. The dependent measure is the amount of fluid consumed.

	Stressed (crowded)	Not stressed (not crowded)
P (alcohol preferring)	40	40
NP (alcohol not preferring)	40	40

Figure 1. Stress and alcohol experimental design.

The specific hypotheses being tested are:

1. The P rats in both crowded and uncrowded conditions will consume more alcohol than the NP rats in the same (crowded or uncrowded) conditions.

2. The strain difference in alcohol consumption will be dramatically greater in the crowded than it will be in the uncrowded conditions.

3. The amount of alcohol consumption in NP rats will not be affected by crowded conditions.

Rats will be tested over a two-month period. In order to get accurate measures of the amount of the alcohol solution consumed by each animal, each rat will taken out of the cage three times per day (once every six hours beginning at 6 A.M.) and put into a standard one-rat cage for a period of one hour where it will be given access to a 10 percent alcohol/water solution and food. Blood samples will be drawn from each rat on Day 1, Day 30, and Day 60 to obtain levels of blood cortisol, adrenaline, and adrenocorticotrophin hormone. These tests will serve as biochemical measures of the stress level of each rat. Such biochemical measures can be used to verify the degree of stress reaction to the crowded conditions by comparing levels between the crowded and uncrowded groups of rats.

After the two-month test period, all rats will be decapitated using a guillotine, and their heads will be immediately frozen in liquid nitrogen. The brain tissue will later be sectioned and subjected to microscopic analysis to determine whether there are any neurological differences between the various groups. The rats cannot be anesthetized before decapitation because anesthetizing the animals could alter the brain tissues.

SIGNIFICANCE

This experiment will explore the effects of stress, here caused by crowding, on alcohol consumption. It will answer the following questions:

1. Does alcohol consumption increase with crowding-induced stress?

2. Is the increase in alcohol consumption among crowded animals greater for P than for NP rats?

3. Are any of the above differences reflected in the animals' brain tissue?

Since previous studies have indicated that the P rats are a good model for human alcoholism, the results of these experiments will help us to determine the roles of stress and genetic factors in the development of alcoholism. This understanding may aid us in designing treatments for alcoholics and those who may be at risk for alcoholism.

Questions for Discussion

You are a member of State University's institutional animal care and use committee (IACUC), which comprises (among others, perhaps) an ethicist, a biologist, a psychologist, a veterinarian, a minister, and a clerical worker from the local community. Recently, Stop the Killing Now (SKN), a radical student animal-rights group at the university, has been putting pressure on the university to stop all animal experimentation. Because of this pressure, the university has asked the committee, which last year approved the psychology laboratory's research on alcoholism and rats, to reconsider that decision. As a member of the IACUC, your task is to decide whether to allow the research to proceed.

1. In applying the regulations and other considerations for evaluating the proposal (see the Case Notes at the end of this chapter), what information will you need to bring to bear on your decision? For instance, how should IACUC members decide the "scientific importance" of a given experiment? On what scientific and social ground is the issue of merit in research decided? On what basis *should* merit be decided?

2. What will you decide about whether to allow the research in this case to proceed? On what basis did you reach your decision?

3. Of what relevance to your decision, if any, is the importance of the proposed research? How did you determine the relevance and significance of this factor?

Part B. The Public Response

The IACUC has approved this experiment, which is currently in its fifth week. Drs. Jones and Smith believe that they are starting to see some very interesting trends in the behavior of the crowded rats. However, SKN has learned that these experiments are taking place on the campus. The members of SKN have decided that both the breeding of alcoholic rats and their use in this research project are cruel and inhumane and should be stopped immediately.

For the past three weeks, SKN has been passing out leaflets and picketing in front of the psychology and administration buildings. The leaflets and signs accuse the researchers of torturing and murdering innocent animals. Last week, they staged a midnight raid on the laboratory to "liberate" the rats and in the process released 55 P and NP rats from the breeding stocks into the wild, destroyed valuable equipment in Dr. Smith's lab, and stole irreplaceable research notes from Dr. Jones's office. Yesterday, they began phoning the researchers at home, calling them hourly throughout the night to accuse them of being murderers. One of these calls was taken by Dr. Jones's nine-year-old daughter, who was so upset by the call that she was unable to attend school today.

The university administration is reluctant to stop the research, citing a history of free scientific inquiry, the fact that the work follows all state and national animal use guidelines, and that valuable knowledge about alcoholism will result. In addition, the generous funding that the alcohol-preferring rat breeding program and research have obtained from federal and private sources is a significant economic benefit to the university. Because the state legislature is contemplating a significant reduction in the university's budget for the next five years, the administration is reluctant to forgo any possible sources of funding.

Questions for Discussion

1. Regardless of your decision regarding whether the research should be conducted, was SKN's conduct ethical?

2. If your IACUC decides that State University should prohibit all further research involving experiments on rats relating to alcoholism, what action, if any, could the researchers take if they want to continue with their research?

■ Case 2: Research on Monkeys: Use or Abuse?

Dr. X, a noted researcher, carries out federally funded studies with monkeys on the effects of destruction of sensory nerves (deafferentation) on limb use.[3] The goal of the research is to ascertain whether persons who have lost the use of a limb due to accidental deafferentation or stroke-induced neural damage can regain use of that limb. Initial findings from Dr. X's experiments suggest that a deafferented monkey can regain the use of a limb if it is forced to use the limb.

The deafferentation procedure is done under sterile surgery and the monkey is under anaesthesia. The surgery involves making an incision in the monkey's arm, cutting the nerve, and closing the incision. The deafferentation procedure was usually done on relatively young monkeys in order to study how brain reorganization is affected as the animal develops. The animal can no longer feel sensations in the limb once the nerve is cut, so no pain is involved in the operation. However, the animal lost the use of that limb, at least temporarily.

The experiments have shown that the monkeys can learn to use a deafferented forelimb if the good one is restrained. This research has potential application to human patients suffering from brain injury, stroke, and other forms of neurological damage who were originally thought to have reached the extent of their abilities to relearn use of deafferented limbs by conventional methods (Miller 1985).

While Dr. X was investigating the mechanisms involved in the recovery of function, Ms. Y, an assistant in Dr. X's laboratory, sent the police several photographs which, she claimed, depicted cruel and inhumane use of the monkeys in the deafferentation experiments. The photographs showed that the monkeys' cages were unclean and that their wounds were unbandaged. The monkeys appeared listless and unkempt.

The police raided the laboratory and arrested Dr. X. He was subsequently charged with more than a dozen counts of cruelty to animals. In addition, his federal funding was suspended.

Dr. X claimed that the photographs had been "staged," that the animals had in fact received humane and gentle treatment, and that leaving the wounds unbandaged was done deliberately for the benefit of the animals, since, if bandaged, they would have further injured themselves by trying to chew off the bandages. (Monkeys tend to put things into their mouths that they find new and interesting, including bandages. Because deafferented monkeys have lost sensation in their bandaged arms, they tend to chew beyond the bandage and may thereby injure themselves. In addition, young monkeys, like humans, tend to harass cripples, thus creating another risk to monkeys that have been deafferented.)

Dr. X further claimed that the unsanitary conditions portrayed in the photographs were misleading: even though the cages were cleaned several times each week, monkeys are naturally dirty. Dr. X pointed out that one of Ms. Y's responsibilities in the laboratory was to clean the monkeys' cages, so that if the animals had been kept in unsanitary conditions it was partly her fault.

As the case unfolds, it is discovered that Ms. Y is a member of Liberate Animals Now (LAN), a radical animal-rights group that has frequently used violent tactics to prevent experiments on animals. In the court proceedings, Dr. X accused Ms. Y of staging the scenes depicted in the photographs. Ms. Y admitted that she chose to work in Dr. X's laboratory with the intention of stopping his research on monkeys as part of a larger LAN scheme to eliminate all research on animals. Although at first she denied staging or doctoring the photographs, she later argued that they were representative of the "type" of treatment that animals received in Dr. X's laboratory.

After several years during which the research was in abeyance (but before the legal proceedings were concluded), a group of researchers requested permission to use the monkeys, which were neither old nor sick, for an additional experiment. They wished to investigate whether the prolonged deprivation of sensory input to specific areas of the brain produced by the deafferentation might change the responses of these areas to sensory stimulation. Electrical potentials would be recorded from the brains of anaesthetized animals when, under anaesthesia, various body parts were stimulated by gentle brushing with a cotton swab or a camel's hair brush. Upon conclusion of the experiments the animals would be euthanized by an overdose of pentobarbital.

LAN sought to block this research, but the court denied LAN's request. The research was then carried out as described above, with surprising results. It was found that the somatosensory area of monkey cortex that had been deprived of its normal input by the deafferentation procedure now responded to stimulation of a portion of the monkey's face, from the chin to the lower jaw. The expanded face representation exceeded the limit of reorganization previously observed as a result of deprivation of sensory input " . . . by an order of magnitude and leaves[s] open the possibility that the limit is even greater" (Pons et al. 1991).

Questions for Discussion

You are a member of the IACUC at Dr. X's institution.

1. Would you have approved Dr. X's proposal before the research began? What is the rationale for your decision? What effect, if any, does the fact that the experimental subjects were monkeys (rather than, say, cats or rats) have on your decision? What restrictions or stipulations, if any, would you have placed on this research? (Review the IACUC Guidelines provided in Case Note 1.)

2. Taking into account Ms. Y's claims that the animals were mistreated and Dr. X's rebuttal, would you vote to terminate the IACUC's authorization of the research?

3. Suppose that LAN has asked the IACUC to hear its emergency petition to block the final experiments involving electrical recording from the brains of the monkeys. Recall that the monkeys must be euthanized upon completion of the experiments. Would you vote to stop these experiments? If these experiments are not to be carried out, what *should* be done with the monkeys?

Case Notes

1. IACUC *Guidelines.*

The deliberations of institutional animal care and use committees take into consideration many factors when reviewing research protocols involving animal research subjects. If you were a member of an IACUC, you would consider:[4]

a. The number of animals to be used and the severity and length of pain to which they will be exposed. (This guideline assumes that the less invasive the procedure, the less ethically problematic the experiment—for example, that experiments measuring only behavior are less problematic than those involving physical manipulation of or harm to the animals' physiology.)

b. The purpose of the experiment, its scientific importance, and the benefits to society expected from the results.

c. The likelihood that the design of the research methodology will achieve the anticipated or desired results.

d. The level of sentience or complexity of the animals involved in the experiments. (This guideline assumes that it is more difficult to justify experiments on more sophisticated animals, such as monkeys, than less sophisticated ones, such as amoebas. In deciding Case 1, you will need to decide whether rats are more like monkeys or amoebas.)

In addition, the following legal requirements must be met:[5]

> (i) Procedures involving animals will avoid or minimize discomfort, distress, and pain to the animals;
>
> (ii) The principal investigator has considered alternatives to procedures that may cause more than momentary or slight pain or distress to the animals;
>
> (iii) The principal investigator has provided written assurance that the activities do not unnecessarily duplicate previous experiments;
>
> (iv) Procedures that may cause more than momentary or slight pain or distress to the animals will:
>
> (A) Be performed with appropriate sedatives, analgesics, or anesthetics . . . ;
>
> (B) Involve, in their planning, consultation with the attending veterinarian or his or her designee;
>
> (C) Not include the use of paralytics without anesthesia;
>
> (v) Animals that would otherwise experience severe or chronic pain or distress that cannot be relieved will be painlessly euthanized at the end of the procedure or, if appropriate, during the procedure;
>
> (vi) The animals' living conditions will be appropriate for their species . . . and contribute to their health and comfort. The housing, feeding, and nonmedical care of the animals will be directed by the attending veterinarian or

other scientist trained and experienced in the proper care, handling, and use of the species being maintained or studied;

(vii) Medical care for animals will be available and provided as necessary by a qualified veterinarian;

(viii) Personnel conducting procedures on the species being maintained or studied will be appropriately qualified and trained in those procedures;

(ix) Activities that involve surgery include appropriate provision for pre-operative and post-operative care of the animals in accordance with established veterinary, medical, and nursing practices . . . [and] using aseptic procedures;

(x) No animal will be used in more than one major operative procedure from which it is allowed to recover . . . ;

(xi) Methods of euthanasia used must be in accordance with [established regulations]. . . .

2. *An Overview of Ethical Approaches to Animal Experimentation*

Several analytical frameworks can be useful in analyzing the use of animals in research. As you review the following summary, note that the various approaches make different assumptions about the moral status of animals. In addition to this brief summary, refer also to Chapter 2, which describes these and other ethical theories in greater detail.

1. The *utilitarian* approach may be roughly summarized as applying a cost-benefit analysis to moral decision making. In focusing on the *consequences* or *outcomes* of actions rather than the principles that underlie them, the person using this approach asks two important questions:

a. What is the good or benefit to be achieved by this experiment using animals?

b. What is the harm or detriment that will occur to the animals that are the subjects of the proposed research?

If animals are viewed as having feelings and awareness, then their suffering must be taken into consideration as one of the harms of using them in research. This harm, as well as any other foreseeable harms, including harms to humans, is then balanced against the benefits that are likely to result from doing the research. Potential benefits to humans as well as to animals, possibly including even those being used in the experiment, are included in this analysis. Only if the benefit outweighs the harm is the experiment considered ethically justifiable. For example, if an experiment on AIDS is not likely to effect a cure for the subjects of the experiment themselves, but may lead to a later cure for AIDS, a utilitarian approach might still consider the experiment to be acceptable.

A difficulty in using this balancing approach is determining in advance whether a proposed research program will actually prove beneficial to human welfare or will result only in a loss of animal life.

2. A *deontological* approach focuses on some criterion other than the con-

sequences or outcomes of actions, such as *principles*. In assessing ethical conduct, the deontologist asks whether the action is right or wrong, regardless of what the consequences are. Here, the interests of animals are viewed in terms of rights rather than the avoidance of suffering. Deontologists are divided about whether animals have a status that would require us to afford them rights.

One way to analyze these cases is to consider whether or not animals have inherent value. Deontologists assume that human beings have inherent value (that is, a worth and entitlement to respect) simply because they are human. (Contrast instrumental value, which is valuing something because of what it is worth economically or how it can be used.) Thus, if A unjustly harms B, A's action is considered to be wrong, at least in part, because it violates the principle of respect for persons. If animals have inherent value, they are also entitled to be treated with respect. But because it is not evident *what* gives something inherent value, the principles of deontology are not self-evidently applicable to animals.

In addition to the utilitarian and deontological approaches, the ethics of animal research has also been considered using other theories, including the following:

3. *Contractarian* approaches, which view moral obligations as founded on voluntary associations between moral agents. Moral agents are generally defined as persons who are able to make decisions for themselves about right and wrong. A requirement of moral agency implies that animals cannot have rights, nor can humans have duties to them, because even if animals are entitled to a moral status, most animals are not moral agents. Feelings of affection or sympathy are not the same as duties.

4. *Kinship* approaches, which argue that we have stronger duties to care for those closest to us (families, relatives, other humans) than for those who are more distant or different, including animals.

5. *Organic unity theory*, which views all organic life as having inherent moral value (Russow 1990: 7), and would give all animals the same status as humans. This approach would especially promote care and respect for animals that are part of an endangered species or ecosystem.

Notes

1. The federal government regulates the handling, feeding, care, transport, and use of pain-relieving drugs for research animals in the Animal Welfare Act, first passed in 1966 and amended several times since. The Health Research Extension Act of 1985 is the other major federal law governing animal research. It requires researchers funded by the Public Health Service to comply with a number of other laws and to report their compliance periodically.

2. The National Institute of Mental Health requires that IACUCs include scientists, a veterinarian, nonscientists, and community members.

3. This case is a simplified account of a real series of events widely reported in the media from 1981 to 1990.

4. These criteria are adapted from Donnelly and Nolan 1990, n. 2.

5. These criteria are quoted from U.S. Department of Agriculture, "Animal Care Rules and Regulations," *Code of Federal Regulations*, Title 9, Part 2, Subpart C, Section 2.31(d), "Institutional Animal Care and Use Committee (IACUC)."

References and Suggestions for Further Reading

Etlin, David. "Director of Research Laboratory Is Guilty of Cruelty to Monkeys." *Baltimore Sun* (November 24, 1981), p. A1. (Reporting on the success of PETA's lawsuit against Dr. Edward Taub, resulting in his conviction on six counts of cruelty to animals for failing to provide veterinary care to the 15 Silver Spring monkeys he used in deafferentation experiments.)

———. "Taub Denies Allegations of Cruelty." *Baltimore Sun* (November 1, 1981), p. 13. (Reporting on the progress of PETA's lawsuit against Dr. Taub.)

———. "House to Hear About Monkeys." *Baltimore Sun* (October 11, 1981), p. 5. (Reporting on House of Representatives subcommittee hearings on PETA's allegations of mistreatment of animals against Dr. Taub.)

Donnelly, Strachan, and Nolan, Kathleen, eds. "Animals, Science, and Ethics." *Hastings Center Report* 20(3): 1–32 (Special Supplement, May/June 1990).

Fraser, Caroline. "The Raid at Silver Spring." *New Yorker* (April 19, 1993), pp. 66–84. (Traces the events and follows up on the fate of the monkeys and Dr. Taub's research. Also describes the origins of PETA and the animal-rights movement in the United States.)

Leary, Warren. "Lawsuit Planned to Halt Killing of Lab Monkeys." *New York Times* (Jan. 18, 1990), p. 11. (Reporting on PETA's lawsuit against Dr. Taub.)

Li, T.-K.; Lumeng, L.; McBride, W. J.; and Murphy, J. M. "Rodent Lines Selected For Factors Affecting Alcohol Consumption." *Alcohol and Alcoholism*, Supp. 1 (1987), pp. 96–99.

———. "Alcoholism: Is It a Model For the Study of Disorders of Mood and Consummatory Behavior?" *Annals of the New York Academy of Sciences* 499: 239–249 (1987).

McCabe, Katie. "Who Will Live, Who Will Die?" *Washingtonian* 21(11): 112–118, 153–157 (August 1986).

Miller, Neal. "The Value of Behavioral Research on Animals." *American Psychologist* 40(4): 423–440 (April 1985).

Murphy, J. M.; McBride, W. J.; Lumeng, L.; and Li, T.-K. "Contents of Mono-Amines in Forebrain Regions of Alcohol-Preferring (P) and Non-Preferring (NP) Lines of Rats." *Pharmacology, Biochemistry & Behavior* 26: 389–392 (1987).

Palca, Joseph. "Famous Monkeys Provide Surprising Results." *Science* 252: 1789–1790 (June 28, 1991). (Reporting on research conducted on the Silver Spring monkeys after the commencement of the lawsuit brought against Dr. Taub.)

Pons, T. P.; Garraghty, P. E.; Ommaya, A. K.; et al. "Massive Cortical Reorganization After Sensory Deafferentation in Adult Macaques." *Science* 252: 1857–1860 (June 28, 1991).

Russow, Lilly-Marlene. "Ethical Theory and the Moral Status of Animals." *Hastings Center Report* 20(3): 4–8 (Special Supplement, May/June 1990).

Stewart, R. B.; Murphy, J. M.; Lumeng, L.; and Li, T.-K. "Differences in Performance on

Anxiety Tests in Rats Genetically Selected for Ethanol Preference." *Society for Neuroscience Abstracts* 15: 61 (1989).

U.S. Department of Agriculture. "Animal Welfare Rules and Regulations." *Code of Federal Regulations* Title 9, Part 1, "Definition of Terms;" Part 2, "Regulations;" and Part 3, "Animal and Plant Health Inspection Service."

10 Research Involving Human Subjects

Psychologists have a duty to treat human subjects ethically. This exercise is designed to introduce ethical issues raised by the involvement of human subjects in psychological experiments. The general issue presented concerns the dimensions of researchers' responsibilities to the subjects who participate in their experiments. The case that follows is a proposal for an experiment examining the relationship between alcohol intoxication (or the belief that one has consumed an alcoholic beverage) and aggressive behavior.

Bioethicists generally agree that there are three principles according to which ethical research involving human subjects is conducted:

1. respect for persons;
2. beneficence; and
3. justice.[1]

Respect for persons means treating others as autonomous agents, or "ends in themselves," having rights and the freedom of self-determination, rather than as means to an end. The principle of respect for persons underlies the requirement that human subjects give their free, voluntary, and informed consent to participate in research. Beneficence toward human subjects means that research should be for the good of the subject (if not directly, then indirectly through benefiting society), and that possible benefits are maximized and risks are minimized. Justice means that benefits and harms are to be distributed fairly.[2]

For this exercise, you and your classmates will play the role of an institutional review board (IRB). IRBs have been set up in universities, hospitals, and other places where research is conducted involving human subjects. Their main task is to ensure that research is conducted in an ethical manner so that the welfare of subjects is protected. As you read the following research proposal, think about whether the criteria of respect for persons, beneficence, and justice have been met.

A "Modest" Research Proposal on Alcohol Experimentation

Purpose

Crime statistics consistently reveal a relationship between aggressive behavior and alcohol consumption. For example, approximately 50 percent of all rapes and 40 percent of murders and cases of child abuse are associated with alcohol intoxication (Eckard et al. 1981). Despite these statistics, the precise mechanisms (psychological or pharmacological) that mediate this relationship are not known. A relatively large body of data indicates that many of alcohol's psychosocial effects are mediated by the person's *belief* (expectancy) that he or she has consumed alcohol rather than by the pharmacological effects of the drug per se.

The purpose of this study is to examine the relative contributions of the pharmacological effects of alcohol and the belief that one has consumed alcohol on aggressive behavior in a highly controlled experimental context. The design of the experiment, outlined below, will enable us to test the following hypotheses:

1. The *belief* that one has consumed alcohol may lead to increased aggression, regardless of whether the individual *actually* has consumed alcohol (referred to as an expectancy effect).
2. Alcohol alone may increase aggression even when the individual does not know that she or he has actually consumed alcohol (a purely pharmacological effect).
3. Some combination of the expectancy and pharmacological effects results in increased levels of aggression.

The experimental design that can address these important questions is referred to as the "balanced-placebo" design, where an alcoholic or nonalcoholic beverage is administered to subjects in one of four different manners (beverage conditions):

1. Subjects are told they will be receiving alcohol (vodka and tonic) and actually receive the alcohol (expectancy and pharmacological effect).
2. Subjects are told they will get an alcoholic beverage and actually receive only tonic (expectancy only).
3. Subjects are told they will receive a tonic and actually get a vodka and tonic (pharmacological effect only).
4. Subjects are told they will receive only a tonic and get only a tonic (neither effect).

Central to this type of design is the successful deception of subjects in conditions 2 and 3. Subjects are not informed of any deception prior to participation. The consent form will indicate that the subject agrees to drink either a vodka and tonic beverage or a tonic water beverage.

PROCEDURES

Sixty-four male nonalcoholic subjects between the ages of 21 and 25 will be recruited for this project through a newspaper advertisement requesting responses from male social drinkers in this age group interested in participating in paid research at State University's Department of Psychology.[3] Each subject calling in response to the advertisement will first be screened on the telephone to determine that:

1. he meets the age requirements;
2. he drinks alcohol regularly, but not in excess of four occasions per week and six drinks per occasion;
3. he has not been arrested for any type of offense, received treatment for alcohol problems, or experienced other alcohol-related problems; and
4. he has not received treatment for any type of psychological disorder.

If a subject fulfills these criteria, he will be scheduled to come to the laboratory and will be told that he must bring at least two pieces of identification verifying his age.

Prior to participation in the actual experiment, all subjects will read and sign a consent form outlining the procedures of the experiment, will have their ages verified (a staff person will examine two pieces of identification), and will complete a standardized questionnaire that will further screen for the presence of alcohol abuse. Subjects will then participate in a hearing test and a test to determine the loudness level at which a specific tone is uncomfortable to the subject. All subjects will be paid $5.00 per hour for their participation. Subjects will be told that they can withdraw from the experiment at any time.

The 64 subjects will be randomly assigned to one of the four beverage conditions of the balanced-placebo design described above. After instructions are given for the task (see below), subjects will receive their drink and be required to consume it within ten minutes. For subjects receiving alcohol, the actual volume (dose) of alcohol given to each subject is determined based on body weight and calculated so as to bring his blood alcohol level up to 0.05 percent. Such a dose is below the legal intoxication limit but enough to make the individual mildly intoxicated. After consuming the alcohol, each subject will undergo an identical testing situation designed to examine aggressive behavior.

AGGRESSION TEST

Subjects will be told that they will be participating in a reaction-time/pain-perception test with a partner-subject in another room. Subjects will be seated in a test room in view of a video camera. They will be told that the video camera is used to allow the partner to monitor the instructions being given to the subject. (The camera actually records the subject for later study by the investigators.) The subject will be instructed that there are actually two experiments being conducted, one that measures reaction time, for which he is a subject, the other measuring pain perception and endurance, for which the other individual (partner) is a subject.

In summary, the instructions state that during the experiment, the partner will present a particular tone to the subject indicating that a reaction-time trial will commence. Five seconds later, a light will flash on a console, after which the subject is to press any one of five buttons (numbered 1 to 5) as fast as possible. A button press will deliver a shock to the partner. (Button 1 is a barely perceptible shock; buttons 2-5 activate increasingly painful, but not harmful, shock levels.) The nonharmful nature of the shocks and the possibility of pressing only shock button 1 will be clearly stated to the subjects. After receiving a shock, the partner will allegedly signal the degree of pain experienced by activating one of five tones of differing loudness levels. (In actuality, the computer generates a tone of a loudness level (1 through 5) that corresponds to the shock intensity level delivered by the subject.) The loudness levels increase, so that tone 5 is five decibels louder than the level previously determined to be uncomfortable for the subject. The subject will also receive feedback by means of five lights on a console. The lights indicate the loudness level of the tone delivered by the partner. The experiment will include 22 reaction-time/pain-perception trials.

After receiving these instructions, the subject will observe a video monitor where he will see instructions being given to the alleged partner. In actual fact, there is no partner; rather, the instructions to the partner are pre-recorded using a bogus partner. The entire reaction-time task will be conducted with the subject interacting with a computer. The subject will not actually deliver shocks to anyone. The video monitor and camera are present only for the purpose of deceiving the subject into thinking that another person in the next room will be receiving the shocks (and for recording the subject for later study by the investigators).

After instructions have been given to the subject and his alleged partner, the video monitor will be turned off and the subject will consume his beverage and have his breath-alcohol content measured with a breathalyzer. Then the aggression test will begin. The measures of aggression will be both the length of time that a subject presses a shock delivery button and the level of shock that the subject chooses.

Following the experiment, each subject will be debriefed. Subjects will be informed that they were deceived, that they did not deliver a shock to anyone, and that their behavior (shock delivery) was completely normal. Subjects will be given the opportunity to express how they feel about having been deceived. In addition, the subjects will be told that deception was necessary in order to determine the answer to the important question of whether the psychological factor of alcohol expectancy influences aggressive behavior.

Questions for Discussion

1. In your opinion, which of the following options best describes the decision that should be reached concerning this proposed experiment?

a. "Accept without revision." (You decide to permit the research to be carried out as described, with no modifications or additional precautions required beyond those already specified in the proposal.)

b. "Tentatively accept contingent upon revision." (You decide to permit the research to be carried out, but with some modifications that you specify.)

c. "Reject with permission to resubmit." (You decide that substantial revisions are necessary in order to meet ethical standards, but the project is not fatally flawed.)

d. "Reject absolutely." (You decide not to grant permission to resubmit, based on the conclusion that the experiment can not be made to comply with ethical standards.)

What reasons can you give to support your decision?

In order to assist you in making your decision, it may be helpful to know that IRBs are often asked to evaluate proposed research according to criteria such as those suggested by the American Psychological Association (APA). These criteria are consistent with the three principles of the Belmont Report (respect for persons, beneficence, and justice), and with federal regulations governing research that involves human subjects:

a. whether the risks are reasonable in relation to the benefits;

b. whether the risks have been minimized as much as possible within the limits of the experimental design;

c. whether subjects have been selected equitably;

d. whether valid informed consent has been obtained;

e. whether privacy of subjects has been protected and confidentiality of data has been maintained;

f. whether the data-gathering and experimental phases of the experiment have been monitored for safety and adequate steps have been taken to protect the subjects from physical or mental discomfort or danger; and

g. whether adequate additional protections have been provided for subjects who are likely to be vulnerable to coercion or undue influence.

Is the experiment described in this case ethical according to these criteria? Why or why not?

2. Is it possible to obtain *informed consent* when the research involves deception?

3. When, if ever, is deception ethically justifiable in experiments involving human subjects?

The APA's *Ethical Principles in the Conduct of Research with Human Participants* requires that before conducting a study that makes the use of concealment or deception necessary, "the investigator has a special responsibility to (1) determine whether the use of such techniques is justified by the study's prospective scientific, educational, or applied value; (2) determine whether alternative procedures are available that do not use concealment or deception; and (3) ensure that the participants are provided with sufficient explanation as soon as possible" (Principle E, 1982). Have these criteria been met in this case? Are they, in your opinion, adequate criteria for justifying research that involves deception?

Notes

1. See *The Belmont Report* (U.S. National Commission for the Protection of Human Subjects of Biomedical and Behavioral Research 1978). See also Beauchamp and Childress 1989, which includes the additional principle of nonmaleficence.

2. For a more extensive discussion of the Belmont Report and these three principles, see Veatch 1987 and Beauchamp and Childress 1989.

3. The study arbitrarily uses all males because of the desire that the subject population be homogeneous.

References

Beauchamp, Tom, and Childress, James F. *Principles of Biomedical Ethics*. 3d ed. New York: Oxford University Press, 1989.

Eckard, Hartford, et al. "Health Hazards Associated with Alcohol Consumption." *Journal of the American Medical Association* 246: 648–666 (1981).

Pihl, R. O.; Smith, Mark; and Farrell, Brian. "Alcohol and Aggression in Men: A Comparison of Brewed and Distilled Beverages." *Journal of Studies on Alcohol* 45: 278–282 (1984).

U.S. Department of Health and Human Services. "Protection of Human Subjects." *Code of Federal Regulations*, Title 45, Part 46 (1990).

U.S. National Commission for the Protection of Human Subjects of Biomedical and Behavioral Research. *The Belmont Report: Ethical Principles and Guidelines for the Protection of Human Subjects of Biomedical and Behavioral Research*. Washington, DC: U.S. Government Printing Office, 1978. Reprinted in *Federal Register* 44: 23192 (April 18, 1979).

Veatch, Robert M. *The Patient as Partner: A Theory of Human-Experimentation Ethics*. Bloomington: Indiana University Press, 1987.

11 Research Involving Human Subjects
The Administration of Alcohol

THE STUDY OF alcohol use, abuse, and dependence is a major national research priority. Alcohol abuse and dependence is associated with devastating personal and social consequences. It is currently the third most prevalent public health problem in the United States. Alcohol use is associated with approximately one-half of all automobile accidents, homicides, rapes, acts of family violence, and child molestations. Costs to the U.S. economy are estimated to exceed $120 billion annually.

To help us understand more fully the factors that influence alcohol consumption and abuse and the factors that may mediate how a particular individual may be affected by alcohol, a large number of studies examine the effects of alcohol by administering alcohol to human subjects in a laboratory setting. The research focuses primarily on the question of whether alcohol actually reduces cardiovascular stress reactivity, thereby acting as a potential reinforcer of drinking behavior in stressful situations. (There is disagreement among researchers as to whether alcohol produces a true physiological stress-reduction effect or whether some subjects experience pronounced stress reactivity when intoxicated.) The secondary purpose of the research is to identify the factors that might cause alcohol to reduce the reaction to stress in some individuals and those factors that might cause alcohol to increase reactions to stress in other persons. Identification of these factors may lead to the identification of characteristics that are related to an increased likelihood of alcohol abuse.

The following cases describe experiments in which alcohol is administered to a subject and various measures of the individual's responsivity to alcohol are measured. All of the cases employ a similar experimental protocol, but differ with respect to (1) the subjects' characteristics and (2) the information collected from the subjects. Each scenario involves the use of "mild" stress induction, a commonly used methodology that poses minimal risk to participating subjects.

The ethical issues presented in the cases in this chapter pertain to the use of experimental procedures that pose a risk of harm to subjects. In working through the cases, pay attention to defining the nature of the (potential) harm, the procedures that might be employed to reduce or eliminate the risk of harm, the reasonableness of the research in view of the potential harms, and the unique conceptual issues presented by the problem of identifying the degree of risk posed by the research.

The cases involve various methods through which a group of investigators plan to recruit participants in an alcohol study. The experimental procedure is first described, followed by each of three recruitment scenarios and questions for discussion.

Methodological Background

A widely used experimental approach to research such as this is to study various groups of individuals whose level of overall risk for alcohol abuse varies. The basic rationale behind such an approach is that the factors that might lead to alcohol abuse, which, in this case, might be related to a greater sensitivity to the reinforcing effects of alcohol, are more likely to be found in groups at greater risk. Individuals are assumed to be at greater risk for alcohol abuse if they possess a characteristic or are a member of a group that is associated with greater prevalence of alcohol abuse than is found in the general population. Described below are some characteristics (e.g., a history of previous alcohol abuse) and groups (e.g., males) associated with increased risk for abuse.

Note both that there are different risk factors as well as varying degrees of risk. In general, the more risk factors an individual possesses, the greater the overall level of risk. For example, one risk factor is being male. Males exhibit two to three times the levels of alcohol abuse as do females.[1] Thus, at any given point in time, the average male has a 7.5 percent chance of developing serious alcohol problems within three months. Another risk factor is age. In general, young adults (between 18 and 30 years of age) have higher prevalence rates of alcohol problems than do older adults. The risk that a younger adult will experience alcohol problems is close to 13 percent (20 percent for men and 6 percent for women). Another risk factor, which appears to be present only in males, is being the child of an alcoholic parent. Sons of alcoholic parents have been estimated to have a risk level upwards of 35 percent. Another risk factor is having a history of alcohol- and drug-related problems. Such individuals have a risk level of approximately 50 percent. In this case, mediating factors relevant to the risk prediction are the length and severity of the earlier problem, as well as amount of time that has elapsed since the earlier period of substance abuse. A person who has had a long history of relatively severe substance abuse in the recent past is at a greater risk for current problems. There are a number of widely used and high-quality substance abuse screening instruments and diagnostic interviews available to researchers conducting drug and alcohol research.

The Experimental Alcohol Administration Procedure (Applicable to Each Recruitment Scenario)

Researchers at a midwestern university are conducting an experiment examining the effects of alcohol on stress. Following recruitment, each subject is scheduled for the test session. Upon arrival for the session, subjects read a description of the experimental procedure, sign a consent form, and have their IDs checked to ensure that they are 21 years of age or older. During the consent process, subjects are provided with a full description of the purpose of the study. Subjects are then assessed for mood (using standard measures), and resting heart rate levels are taken. A stress-inducement procedure is administered, during which heart rate activities continue to be recorded. Subjects then rest for a period of time until they are completely relaxed (i.e., their heart rate returns to pre-stress levels). Subjects then drink enough alcohol to bring their blood alcohol level up to 0.10 percent, the legal limit for intoxication in most states. The stress-inducement procedure is administered once again, and subjects' reaction is recorded. After the testing, subjects remain in the laboratory until their blood alcohol level drops to less than 0.02 percent.

Recruitment Scenario 1

Subjects are recruited for the study through an advertisement placed in the university's student newspaper. The ad reads as follows: "Volunteers needed for an alcohol research project. Must be over 21 years of age. Call 555-2121." When prospective subjects call, they are asked their age and whether they have ever had any problems with alcohol. If a subject reports being over 21 years old and as having had no prior problems with alcohol, he or she is scheduled to come into the laboratory for the experiment. The experiment consists of the Alcohol Administration Procedure, described above.

Questions for Discussion

1. Does this recruitment method present any ethical problems?

2. Suppose the investigators decide to offer a monetary reward for participating in the research. Would such a reward be relevant to your analysis of Question 1?

3. Suppose the study excluded anyone who has never consumed alcohol. Would this restriction affect your analysis of Question 1?

4. Are there any potential risks of harm inherent in the administration procedure that should be disclosed during the consent process? Are there risks of harm that make the study unacceptable, even if the risks are disclosed? If so, is there any way to reduce or eliminate these risks?

5. Are there any gender-dependent risks that need to be considered (i.e., are there risks for men that don't exist for women or vice versa)?

6. How should the consent form be worded?

7. Suppose the study reveals that the subject is an alcoholic or is at risk for alcohol abuse. Does the investigator have an obligation to inform the subject? Suppose that this information is revealed at the screening stage, so that the subject is ineligible to participate. What should the subject be told?

Recruitment Scenario 2

The researchers now recruit persons who are the offspring of alcoholic parents through various groups such as ACOA (Adult Children of Alcoholics) and AL-ANON (family members of alcoholics). A postcard is sent to members of the local chapters of these two organizations soliciting their participation. The postcard reads as follows: "We are seeking to recruit volunteers to participate in an alcohol research project. Participants must be over 21 years of age. For more information, please call 555-2121."

As in Recruitment Scenario 1, when prospective subjects call, they are asked their age and whether they have ever had any problems with alcohol. If a subject reports being over 21 years old and as having had no prior problems with alcohol, he or she is scheduled to come into the laboratory for the experiment. The experiment consists of the Alcohol Administration Procedure, described above. The only difference between the subjects recruited in this scenario and those recruited for Scenario 1, is that these subjects have a higher risk of developing alcohol problems in the future (statistically speaking), by virtue of being a member of a group with a higher prevalence of alcohol problems.

For this scenario, more extensive interviews are conducted prior to admitting subjects to the study. Each subject is interviewed before the alcohol administration procedure begins to obtain detailed information about the history of alcohol use and abuse, both their own history and that of their family members. In addition, subjects are asked to provide detailed personal information about themselves and their family members.

Questions for Discussion

1. Does this recruitment method present any ethical problems? Explain. How does this scenario differ from Recruitment Scenario 1? Is it more or less ethically problematic? Why?

2. If any additional risks are presented by this experiment as compared with the previous one, what might be done to reduce these risks?

3. Would your assessment of the case differ if the subjects had no history of alcohol abuse, but reported recreational use of other mood altering drugs? Why? What are the important issues to consider here?

4. Suppose the subjects, who are offspring of alcoholics, had a history of drug abuse (as opposed to recreational drug use) or alcohol abuse as well. What ethical issues are presented by including them in the study?

5. What types of conceptual issues may cloud the ethical decision making process about the inclusion of the kinds of subjects described in these cases (i.e., no personal history of alcohol problems, family history of alcohol problems but no personal history, family history and personal history, personal history of drug abuse, recreational drug use, drug abuse)?

6. Suppose the study reveals that the subject is an alcoholic or appears to be in the early stages of alcohol abuse. Does the investigator have an obligation to inform the subject? Does the investigator have any obligations to the subject's family members to communicate to them information learned through the study that may have implications for their health or welfare?

7. Suppose that you are a member of the university's human subjects committee and are reviewing the investigator's protocol. What stipulations might you make concerning the consent process, screening, and debriefing procedures?

Recruitment Scenario 3

The same researchers want to conduct the same experiment with alcoholic subjects. Subjects are recruited using advertisements in a local community newspaper. The ad reads as follows: "Heavy drinkers needed to volunteer for alcohol research studies. Call 555-2121."

Questions for Discussion

1. Does this recruitment method present any ethical problems? Explain. How does this scenario differ from Recruitment Scenarios 1 and 2? Is it more or less ethically problematic? Why?

2. What are the potential risks, if any, inherent in this study? What might be done to reduce these risks? Should such a study be done? Why or why not?

3. Suppose the investigators decide to offer a monetary incentive to participate in the research. Would such a payment be relevant to your analysis of Question 1? Explain.

Recruitment Scenario 4

Suppose that previously conducted research indicates that certain aspects of an alcoholic's response to alcohol may predict his or her response to different pharmacological and physiological treatments for alcohol dependence. A group of investigators want to take this research further by pinning down the specific predictors.

Instead of advertising for subjects, the investigators team up with several alcohol treatment centers and recruit subjects as they come in for treatment. Recruitment into the study takes place before enrollment in the treatment center. There is no screening procedure; persons seeking treatment for alcohol dependence are presumed to qualify for participation. Persons who agree to participate are tested using the Alcohol Administration Procedure described above and are then put on different treatment regimens, depending on the nature of their physiological response to the alcohol administration. Treatment is provided by the treatment center.

Questions for Discussion

1. Does this recruitment method present any ethical problems? Explain. How does this scenario differ from Recruitment Scenarios 1, 2, and 3? Is it more or less ethically problematic? Why?

2. Recall that the results of the Alcohol Administration Procedure are hypothesized to be predictors of successful treatment, but that they have not yet been proven to be so. According to the study design, treatment decisions will be made based on this hypothesis. Is this fact important to a decision whether to allow the study to go forward? Would it make a difference to you if the subject's response to the alcohol administration procedure actually did predict which treatment would be successful?

3. When subjects agree to participate in the study, to what are they giving their consent? The alcohol administration procedure? Treatment for alcohol dependency? What happens if a subject changes his or her mind about receiving treatment at the clinic?

4. Suppose the researchers just wanted to examine the nature of the subjects' reaction to alcohol, but there was no relationship between the research and application of the findings to treatment for alcoholism. Are there any ethical concerns involved in administering alcohol to alcohol dependent individuals for research purposes? Does the proposed recruitment method (recruiting subjects at an alcohol treatment clinic) raise ethical concerns? Explain.

Note

1. The prevalence of alcoholism in women at any given point in time is approximately 2.5 percent, while men have a prevalence rate of about 7–8 percent. The rate of alcoholism in the general population is about 5 percent.

References

U.S. Department of Health and Human Services. National Institutes of Health. Office for Protection from Research Risks. *Protecting Human Research Subjects: Institutional Re-*

view Board Guidebook. Washington, DC: Government Printing Office, 1993, pp. 5–64 to 5–70.

U.S. Department of Health and Human Services. Public Health Service. National Institute on Alcohol Abuse and Alcoholism. National Advisory Council on Alcohol Abuse and Alcoholism. "Recommended Council Guidelines on Ethyl Alcohol Administration in Human Experimentation." Revised June 1989.

12 The Ethics of Deception in Research

ONE FUNDAMENTAL PRINCIPLE underlying the ethical conduct of research involving human subjects is respect for persons. This principle is the basis for the requirement that prior to becoming research subjects, prospective participants must give their free, voluntary, and informed consent to participation. As part of providing consent, subjects must be informed of the purposes of the research, the risks and benefits of participating, and various other kinds of information.[1]

Psychological research, however, sometimes includes elements of deception. That is, subjects are not given an opportunity to provide their informed consent to participation before the investigator collects the data. Examples of deception in research include covert observation (where the subject is unaware that research is being conducted and the investigator merely observes the subjects' activities), participant observation (where the subject is unaware that research is being conducted and the investigator interacts with the subject), and research in which the subject knows that he or she is participating in research but the nature of the research is not fully revealed to the subject prior to the agreement to participate.

Legitimate uses of deceptive research practices are situations in which the research cannot be conducted unless the subject is kept in the dark about the purposes of the research. The federal regulations governing research involving human subjects allow deceptive research to be conducted so long as:

1. the research involves no more than minimal risk to the subjects;
2. the rights and welfare of the subjects will not be affected;
3. the research cannot practicably be carried out without the deception; and
4. where appropriate, subjects are provided pertinent information about the research after participating in the research.[2]

In addition, when approving research involving human subjects, institutional review boards (IRBs) consider not only the legitimacy of the consent process the investigator proposes to use, but also the acceptability of the risks to which subjects will be exposed.[3] In reviewing the acceptability of risks, the IRB looks to see whether risks to subjects are minimized (by using procedures that are consistent with sound research design and that do not unnecessarily expose subjects to risk), and whether risks to subjects are reasonable in relation to any anticipated benefits to subjects. This latter criterion (often referred to as the risk/benefit ratio) includes an evaluation by the IRB of the importance of the knowledge that may reasonably be expected to result from the research.

We present here a series of cases designed to explore the ethical questions raised by the conduct of research that includes elements of deception. The cases consist of descriptions of a group of social psychology experiments that were conducted in the late 1960s.[4] All but one are concerned with the conditions under which subjects will act responsibly in a social situation. (The exception examines whether or not a subject will interact with strangers in a playful activity.) When considering these cases, readers should note that the studies were conducted nearly 20 years before the regulations on human subjects were adopted.

■ Case 1: Lost in the Subway

This study took place in the subways of New York City. A situation was prearranged between two experimenters where one experimenter, E-1, posed as a bystander, and the other, E-2, as an individual who was lost in the subway. Both staged a scene in front of an unwitting subway rider, S, who was randomly selected as a subject. E-1 stood or sat near S. E-2, in the guise of looking like he was lost, approached and asked S whether the subway was going uptown or downtown. E-1 cut in and gave the wrong answer—if the subway was going uptown, he replied "downtown" and vice versa. The dilemma for the subject (S) was whether he or she should give the right information, correcting the bystander. Variations of the situation were tried where both S and E-1 were asked (rather than S alone), or E-1 was addressed directly. The dependent measure was whether the subject (S) actually corrected E-1. (Adapted from Latane and Darley 1970: 21–22.)

Questions for Discussion

1. Does this experiment present any ethical problems? If so, what are they? Would your answer be the same if the experimenters had waited in the station and recorded actual situations of people asking directions?

2. Should the experimenter be required to obtain informed consent from the subjects? If so, is there any way of obtaining consent?

3. Should the subject be debriefed following the experiment? If yes, how?

4. Are the subject's rights being violated? If so, how?

5. Might S have been harmed as a result of being chosen as a subject? If so, how?

6. Is the social importance of the experiment relevant to your answers? If so, explain.

Case 2: To Frisbee or Not to Frisbee

This study was concerned with assessing the factors that might promote interaction among previously unorganized groups or strangers in public places. The following situation was staged: A girl sat on a bench in the waiting room at Grand Central Station. Soon another girl sat on a bench facing her. They pretended to recognize each other and began a conversation. One girl had been shopping and announced that she had just bought a Frisbee. The other girl asked to see it, and the first girl threw it to her. They then began to toss it back and forth. Apparently by accident, the Frisbee was thrown to a third person; the reaction of the third person (a confederate or plant of the researchers), was the independent variable of the study. The third person either enthusiastically joined in throwing the Frisbee or accused the two girls of being childish and dangerous and kicked the Frisbee back across the gap. Whichever of these two variations occurred, the two girls continued throwing the Frisbee back and forth and eventually threw it to one of the real bystanders (an actual experimental subject) seated on the benches. They continued this until all the bystanders on the two facing benches had been tried. A bystander was counted as participating in the activity if he or she returned the Frisbee at least twice. The percentage of bystanders who joined in the Frisbee fest was the dependent measure of the study. (Adapted from Latane and Darley 1970: 25.)

Questions for Discussion

1. Does this experiment present any ethical problems? If so, what are they? If the researchers observed and recorded an actual situation of someone trying to involve bystanders in a game of Frisbee, would your answer be the same? Is that approach to collecting this kind of data feasible?

2. Should the experimenter be required to obtain informed consent from the bystander subjects? If so, is there any way of obtaining consent?

3. Should the subjects be debriefed following the experiment? If yes, how?

4. Are the subjects' rights being violated? If so, how?

5. Could any harm could come to any of the bystanders as a result of the attempt to draw them into the game?

6. Is this experiment more or less problematic than in Case 1 in the subway? Why?

7. Is the social importance of the experiment relevant to your answer? If so, explain.

■ Case 3: The Lady in Distress

In this study, male subjects were asked to participate in a questionnaire survey, purportedly being conducted by the Consumer Testing Bureau, a market research organization "interested in testing the market appeal of a number of adult games and puzzles" in exchange for a modest sum of money. As they filled out the questionnaires, they were exposed to an emergency. They were tested alone, with a friend, or with a stranger (co-subjects), and their responses were observed.

Upon arrival for his appointment, the subject was met by an attractive and vivacious young woman (secretary) who showed him to the testing room and gave him the questionnaire to fill out. While he answered the questionnaire, the secretary said that she had a few things to do next door in her office, but would return in ten or 15 minutes to give further instructions. The office door was open and easily accessible from the testing room. While the subject (and co-subjects) worked on their questionnaires they heard the secretary moving around in the next office and opening and closing drawers. After a couple of minutes, they heard a loud crash and a scream as the chair fell over. "Oh, my God, my foot . . . I . . . I . . . can't move . . . it. Oh, my ankle. I . . . can't . . . can't . . . get . . . this thing off . . . me." The secretary moaned for about a minute longer, getting gradually more subdued and controlled. This whole sequence was pre-recorded on high fidelity tape, but the subject (and co-subjects), next door, had no way of knowing that. The main dependent variable of the study was the type of response made to the emergency and the length of time before that response was made. (Adapted from Latane and Darley 1970: 57–58.)

Questions for Discussion

1. Does this experiment present any ethical problems? If so, what are they? If the experimenters had merely observed situations of bystanders' response to situations involving the distress of others, would your answer be the same? Is that approach feasible?

2. Should the experimenter be required to obtain informed consent from the subjects concerning the deception? If so, is there any way of obtaining consent?

3. Should the subjects be debriefed following the experiment? If so, how? What should be said to subjects? Should there be different debriefings for those who did and those who did not come to the woman's aid?

4. Are the subjects' rights being violated? If so, how?

5. Could the subjects be harmed as a result of their inclusion in the study? If so, how?

6. Is this experiment any more or less problematic than Case 1 (in the subway) or Case 2 (with the Frisbee)? Why?

7. Is the social importance of the research relevant to your answer? If so, explain.

■ Case 4: The Hand in the Till

In this experiment, male undergraduates witnessed a (staged) theft while waiting for an interview. In one condition, each subject was the sole witness; in another, two subjects were present. Male college freshmen were asked to volunteer to participate in an interview for a modest sum of money. After arriving for their interviews, the subjects were greeted by an attractive female receptionist and directed toward a waiting room. Among the subjects was a short, clean-cut, conservatively dressed student. This participant (C) was an confederate of the experimenter.

All of the subjects were told that they would be individually interviewed by a team of experts from the Institute for the Study of Consumer Practices on the reactions of college students to the urban environment of New York City. They were also told that since the interviews were running behind schedule, they would be paid in advance.

When paying the subjects, the receptionist pulled several large and small bills out of an envelope in full view of all subjects. To emphasize the presence of the large amount of money in the envelope, she asked the subjects if anyone had change for a fifty. After paying the subjects, she put the remainder of the money (between 150 and 200 dollars) back in the envelope and placed the envelope on top of the desk. Shortly afterward, she left the room to speak to an interviewer.

Just after the receptionist left the room, C walked over to the desk and pretended to fumble with a magazine lying on top of the desk. Seemingly trying to hide his actions but in full view of the other, real, subject(s), C then took the cash from the envelope, picked up the magazine, and returned to his seat. He did not say anything. If a subject questioned him about taking the money, he either ignored the comment, continuing to leaf through his magazine, or innocently answered, "I don't know what you are talking about."

A couple of minutes later, the receptionist returned. At this point in time, the subjects could report the crime and confront C directly. After about half a minute, the receptionist sent C to his "interview." The subjects now had an opportunity to report the theft without directly confronting C.

Finally, each subject was called into his interview. If the subject did not tell the interviewer then, the deceptions and purposes of the experiment were explained. The dependent variable of interest was the proportion of subjects reporting the theft spontaneously to the receptionist or the interviewer. (Adapted from Latane and Darley 1970: 70–71.)

Questions for Discussion

1. Does this experiment present any ethical problems? If so, what are they? Can this experiment be conducted without the use of deception? Is there an alternative research design that would be less problematic (assuming you find it problematic)?

2. Should the experimenter be required to obtain informed consent from the subjects concerning the deception? If so, is there any way to obtain consent?

3. Should the subjects be debriefed? If so, how? What should be said to a subject who did not reveal the theft to the experimenters at any point when he had the opportunity to do so?

4. Are the subjects' rights being violated? If so, how?

5. Could participating in this project as an actual subject pose any harm to the individual? If so, what would that harm be?

6. Is this experiment any more or less problematic than Case 1 (in the subway), Case 2 (with the Frisbee), or Case 3 ("The Lady in Distress")? Why?

7. Is the social importance of the experiment relevant to your answer? If so, explain.

■ Case 5: The Stolen Beer

In this study, two experimenters posed as robbers to conduct a field study examining factors that affect the likelihood that individuals will report crimes they have witnessed. The experimenters were husky young men who would enter a corner store singly or in a pair when some unwitting customers (subjects) were present. They would ask the cashier at the checkout counter: "What is the most expensive imported beer that you carry?" The cashier, in cahoots with the experimenters, would reply, "Heineken. I'll go back and see how much we have." Leaving the experimenters in the front of the store, the cashier would disappear into the rear to look for the Heineken. After waiting for a minute, the experimenters would pick up a case of beer near the front of the store and, remarking out loud, "They'll never miss this," walk out of the front door. The robberies were always staged when there were either one or two potential subjects in the store, and the timing was arranged so that one or both subjects would be at the checkout counter at the time when the experimenters entered. Although occasionally the two subjects had come in together, more usually they were strangers. When the cashier returned from the rear of the store, he came back to the checkout counter and resumed waiting on the subjects at the counter. After a minute, if nobody had spontaneously mentioned the theft, he casually inquired, "Hey, what happened to that man (those men) who was (were) in here? Did you see him (them) leave?" The dependent variables of interest were whether the subjects reported the stolen case of beer, and the time taken to do so.[5] (Adapted from Latane and Darley 1970: 75–76.)

Questions for Discussion

1. Does this experiment present any ethical problems? If so, what are they? Is there an alternative research design that would be less problematic (assuming you find it problematic)?

2. Should the experimenter be required to obtain informed consent from the subjects prior to the experiment? If so, is there any way to obtain consent without biasing the data?

3. Should the subjects be debriefed? If so, how? What should be said to a subject who did not reveal the theft to the clerk at any point of opportunity?

4. Are the subjects' rights being violated? If so, how?

5. Could any harm come to the subjects of this study? If so, what?

6. Is this experiment more or less problematic than the experiment in Case 1 (in the subway); Case 2 (with the Frisbee), Case 3 ("The Lady in Distress"), or Case 4 ("The Hand in the Till")? Why?

Notes

1. U.S. Department of Health and Human Services, "Protection of Human Subjects," *Code of Federal Regulations*, Title 45, Part 46, Section 116 (informed consent).

2. Ibid., 116 (d).

3. Ibid., 111(a) (1) and (2).

4. All of the cases are excerpted from a report of a series of experiments contained in Latane and Darley 1970: 21–22, 25, 57–58, 70–71, 75–76. Adapted by permission of Prentice-Hall, Inc., Englewood Cliffs, NJ.

5. Since the checkout counter was 20 feet from the front door, there were no direct attempts to stop the robberies.

References

Latane, Bibb, and Darley, John M. *The Unresponsive Bystander: Why Doesn't He Help?* Englewood Cliffs, N.J.: Prentice-Hall Publishing Co., 1970.

U.S. Department of Health and Human Services. "Protection of Human Subjects." *Code of Federal Regulations*, Title 45, Part 46.

13 Misconduct in Science

RECENT INCIDENTS OF fraudulent behavior by scientists have drawn attention to the problem of misconduct in science. Misconduct has been defined as "fabrication, falsification, plagiarism, or other practices that seriously deviate from those that are commonly accepted within the scientific community for proposing, conducting, or reporting research" (U.S. Department of Health and Human Services 1989: 32449).

Universities that receive federal money are now obligated to establish procedures for dealing with alleged misconduct: they must ensure that the misconduct is reported, that it is evaluated by an appropriate body, and that, if the charge of misconduct is substantiated, appropriate punishments or remedial measures are carried out. Depending on the severity of the infraction, the punishment for misconduct can be a reprimand, withdrawal of grant support, or, in the most serious cases, termination of employment or even criminal prosecution.

Science is cumulative and ultimately self-correcting. A research investigation typically builds on prior research: it corrects, modifies, and extends previous research findings and, in turn, serves as a basis for subsequent research. Obviously, this cumulative process can work well only if the research is carried out and reported with scrupulous honesty. Dishonesty in research is intolerable because it subverts the scientific enterprise.

The definition of scientific misconduct cited above goes on to point out that scientific misconduct " . . . does not include honest error or honest differences in interpretations or judgements of data" (U.S. Department of Health and Human Services 1989: 32449). Thus, a "wrong" or erroneous result does not per se constitute actionable misconduct. We should not want it otherwise, because, in the long run, scientific results are almost always refuted, reinterpreted, or otherwise modified by later research. Scientists should not have to risk charges of misconduct when they venture into new and uncharted areas of research where errors and misinterpretations are likely.

Somehow, then, we must distinguish between misconduct, which always warrants severe sanction, and behaviors which may be careless, inappropriate, or even unethical but do not constitute misconduct; between dishonest and honest errors in procedure, observation, or judgement; and between fraud and minor deviations from strict honesty in the reporting of experimental results.

Issues of research ethics intrude into every stage of even the most routine research project. Sometimes the ethically appropriate course of action is obvious. More often there is no unequivocally "correct" resolution and the ethical decision is arrived at from considerations of intellectual honesty and what the "consumer" of the research (the reader of the publication or the sponsor of the research) has a right to know about the research. The lapses that occur are typically not such egregious ethical violations as fabrication or deliberate falsification of data. More commonly they are deviations from strict honesty, such as misleading reporting, omission of inconsistent data, failure to maintain procedural standards, or bias in the selection of references.

To illustrate the ethical issues associated with even a rather routine experimental investigation, the successive stages of a research project are presented here. Some stages are described as they might appear in a journal or progress report. Ethical issues that might arise concerning the conduct of the research are interpolated. Some of the investigator's actions described below may constitute misconduct; some may be undesirable or unethical, but are not misconduct; and still others may be entirely appropriate or unobjectionable. What is *your* judgment? If you were to observe these actions, what should you do? Bear in mind that to accuse someone of unethical behavior can damage his or her career and reputation and, possibly, your own. Nevertheless, if you do observe scientific misconduct, you are obliged to take appropriate action.

Part A

Professor X directs a laboratory that studies learning and memory in animals. He obtained funds from a drug company for an investigation of the effects on memory in rats of a patented compound that the company proposes to market as a "memory facilitator" in elderly patients.

1. Professor X's agreement with the drug company states that, while he may publish any data obtained with this compound, he will keep the company informed of the results of his studies and will send the company pre-publication copies of any manuscripts that mention the drug. Are the drug company's stipulations objectionable?

2. Should Professor X accept the funds if the drug company requires that he obtain the company's approval before publishing any data obtained with the compound? Suppose the grantor was the Defense Department, and it stipulated that the results were to be kept secret?

3. Professor X owned 100 shares of stock in the drug company at the time he received the grant. Should he have rejected the grant because of a conflict of interest?

4. Suppose that Professor X is an employee of the drug company rather than a professor, and that he carries out the research in the company's labora-

tories. Under these circumstances, is it appropriate for the company to require its approval prior to publication? For the company to require that the data remain secret?

Part B

Professor X assigned the drug project to J, an advanced graduate student working in Professor X's laboratory. J used a two-lever delayed descrimination test situation that he had developed on his own for measuring the effect of treatments on memory. Rats deprived of water for 24 hours prior to each test session were trained to respond differentially to two stimuli. They received a drop of water when they pressed one lever following presentation of one of the stimuli or if they pressed the other lever following presentation of the other stimulus. Varying time delays were then interposed between the time of stimulus presentation and the opportunity to press a lever. When their performance in this preliminary training procedure had become stable, the animals received intraperitoneal injections of the putative memory-enhancing drug prior to the experimental test sessions. J anticipated that the drug would increase accuracy of performance in this procedure, especially at the longer delays. At the conclusion of the drug series, the experimental animals and nondrugged controls were decapitated, and sections from their brains were examined histologically to see if the drug treatment had altered concentrations of selected neurotransmitters.

J's experimental procedures were approved by his university's committee on animal care and use. However, in order to spare their sensibilities, he did not tell the committee that the animals would have to be decapitated without first being anaesthetized. Does J's omission constitute misconduct?

Part C

In reporting on the research, J stated that he "trained animals on the procedure at the same hour each day, six days a week for five weeks, or until all animals had reached 85 percent accuracy."

The training procedure did not really go as smoothly as the above statement implies. One animal performed poorly for several days and then died. Two animals failed to perform at better than 55 percent accuracy and were therefore discarded. Several times J started the testing sessions an hour or more late, and on two Saturdays he went to a football game instead of testing the animals.

J did not mention any of these irregularities in the report that he prepared for publication. Should he have described any or all of these irregularities in his manuscript?

Part D

In reporting on the research, J stated that he "administered the drug to the animals before test sessions. Each animal received a range of doses of the test drug on different days as well as control injections of the vehicle. The injections were administered in different orders according to an experimental protocol established in advance of the start of the experiment."

1. Once again, the description in the experimental report was less than accurate. The protocol could not be followed exactly. Several animals contracted a respiratory infection midway through the drug series and were sidelined temporarily until they recovered. The computer that controlled the experiments failed twice during drug test sessions, and these test sessions were rescheduled. Which, if any, of these untoward events should have been mentioned in the experimental report?

2. Once, when J was about to inject a drug, he noticed that the rat appeared a bit sickly: its breathing was heavy, and there was exudate around its nose. Another time, after completing a drug injection, he noticed a drop of drug solution on the rat's fur adjacent to the injection site. In both instances, he decided to go ahead with the drug test run and defer decision on recording the results until after the test session. He reasoned that if the data obtained were consistent with previous data, then these irregularities were probably inconsequential, and he would record the results; otherwise he would reject the data. Is J's decision rule acceptable? If not, what *should* be the rule concerning data collection when experimental conditions are "not quite right?"

Part E

The meetings of the Society were coming up. Although the drug series and data analysis had not been completed, J was confident that he could finish the study before the meetings and that he knew what the final results would be. Accordingly, he submitted an abstract describing the expected results for presentation at a poster session. The abstract was accepted and printed in the Proceedings of the Society.

As sometimes happens, J did *not* complete the study before the meetings. Furthermore, the data that J had collected by the time of his poster presentation did not support the conclusions that he had presented in his abstract. Was J's submission of his abstract unethical? Under the circumstances, what action should he take in regard to his poster presentation? What should he do about the abstract?

Part F

The investigation was finally complete, and the data were analyzed. The results were not at all what J had expected when he began the investigation. Most of the animals in the experimental group that received the drug showed distinctly poorer performance than a control group that received only vehicle. Two animals in the experimental group showed no drug effects at all.

1. J decided to eliminate from the data analysis the results from the two rats that did not respond to the drug since their results were so different from those obtained with the other rats. He speculated that these two rats had not been drugged properly, or that their metabolism was peculiar. Is J's decision acceptable? Is it ever acceptable to eliminate data?

2. You will recall that when he began the experiment, J had expected that the animals receiving the drug would perform better than the nondrug controls; in fact, they performed much worse. As he was preparing his results for publication, J examined research that others had performed with the drug and found reasons why the drug may have interfered with performance. Would it be appropriate for J to revise the "hypothesis" to be presented in the introduction to his experimental report so that it would "predict" the results which were actually obtained?

3. The drug company was not enthusiastic about J's manuscript, since it obviously did not support the therapeutic claims made for the drug. If J were to publish the results, he might not receive further grants. Assuming that J's experimental work was scientifically sound, is he *obliged* to publish?

Part G

J prepared a manuscript describing his experimental study and submitted it for publication. Professor X was listed as coauthor.

1. Professor X demanded to be listed as coauthor of the paper. J agreed to do so because Professor X was director of the laboratory and had obtained the grant that supported the research. Besides, Professor X's prestigious name on the manuscript could facilitate its acceptance for publication and increase the visibility of the manuscript once it was published. On the other hand, J had planned and carried out the experiment entirely on his own; Professor X's sole contribution was to read the final manuscript and make a few grammatical corrections. In fact, Professor X had been on sabbatical in Tuamotu while most of the research was being carried out. Should Professor X have been listed as coauthor? What responsibilities does Professor X—as laboratory director and coauthor of the paper—incur for the ethical conduct and accurate reporting of J's research?

2. In his manuscript, J cited several papers that he could not find in the library, copying the reference citations from a previously published paper on the same drug. Does this practice constitute misconduct?

3. In order to enhance his vita, J prepared two different versions of his research report, gave them different titles, and submitted each to a different journal. He also arranged with his fellow graduate student, B, to list B's name as coauthor on one of the manuscripts, in return for B having listed J as a coauthor on a manuscript last year. These practices are unquestionably sleazy; are they misconduct?

Reference

U.S. Department of Health and Human Services. Public Health Service. "Responsibilities of Awardee and Applicant Institutions for Dealing with and Reporting Possible Misconduct in Science." *Federal Register* 54(151): 32446–32451 (August 8, 1989).

14 Science and Coercion

We live in a time that has seen a sustained increase in public concern over coercion and victimization of the powerless by the powerful. Because science is associated with power, some of this concern has focused on the scientific research community. Public interest in controlling the power of science and scientists to coerce and victimize has led to a marked increase in oversight mechanisms, including the establishment of ethics committees in universities, government, and professional societies, all charged with determining the appropriateness of the conduct of scientific inquiry.

The cases in this chapter deal with psychological research activities that may or may not be considered coercive. They should be viewed as examples of the kinds of questions being raised before university and professional ethics committees, particularly institutional review boards (IRBs). IRBs are committees that review and approve research involving human subjects, particularly research funded by the federal government.

Before working with the cases, some background may be useful (see also chapters 10, 11, and 12). One of the basic principles underlying research involving human subjects is the principle of respect for persons. Respect for persons involves a recognition of the personal dignity and autonomy of individuals, as well as special protection for persons with diminished autonomy. The principle of respect for persons underlies one of the fundamental prerequisites for conducting research involving humans: obtaining the free, voluntary, and informed consent of participants. The federal regulations governing the involvement of human subjects in research require that the informed consent of subjects be obtained and that the decision of an individual to consent to participation be free of any coercion (i.e., power or force brought to bear on prospective subjects) or undue influence (i.e., inappropriate financial or other incentives offered to persons who agree to become subjects).[1] Ensuring that recruitment practices are free from coercive or unduly influential elements is one of the responsibilities with which IRBs are charged; it is a key issue in the cases presented in this chapter.

Ethical review of research is another issue raised by the cases. Federal law requires that all federally funded research involving human subjects be reviewed and approved by an IRB before going forward. Another institutional body, the American Psychological Association (APA), is the primary credentialing body for the discipline of psychology and is also concerned with the ethical behavior of both academic and clinical psychologists. Complaints against psychologists claiming ethical breaches may be directed to the APA's Ethics Committee. This committee has the power to censure psychologists and terminate their membership in the association.

The decisions made by these bodies are quite important to the careers of individual scientists. For example, a vote by the APA's ethics committee to censure a psychologist can result in revocation of a therapist's license to practice. The decisions of both IRBs and the ethics committees of professional associations also have a marked influence on the course of science because of their effect on the types and amounts of experimentation that take place.

Research involving students has been considered problematic because of the potential for coercion inherent in the researcher-student relationship.[2] The topic of student participation in research is among the subjects examined in the first four cases. The fifth case explores the question of control over research and its impact on the persons or communities who are the subject of the research.

■ Case 1

All students taking introductory psychology at a major midwestern university are required either to serve in three hours of experiments or to write up a short paper reviewing the experimental methodology in a set of published studies. If a student chooses to participate in experiments rather than write the paper, he or she must sign up for the available experiments as they are posted over the course of the semester. If a student fails to complete the required experiments or the alternative review paper, he or she is given an "incomplete" in the course until the requirement is satisfied.

The stated goal of the experiment-participation requirement is pedagogical. Participation in experiments is intended to acquaint students firsthand with the nature of psychological research. To this end, the students are debriefed and provided with a complete explanation of the research hypotheses and methodology following each experiment in which they participate. In addition to meeting the stated pedagogical goal, the requirement also provides a pool of subjects from which faculty and graduate students can draw for their research.

Student Y missed her first experiment because her car had a flat tire on the way to campus. When she reported to the department office that she had missed the experiment, the overworked and unsympathetic secretary reminded Y that she not only had to make up the missed hour but also had to perform an extra hour of research. Later that day, prodded by irritation, a sense of injustice, the cost of a new tire, and her law student roommates, Y wrote the Psychology Department a letter stating that she refused to perform *any* of the required experiments on the grounds that the department was coercing her into subsidizing the careers of experimental psychologists by using her in place of paid subjects. She then lodged a complaint with the university's human subjects institutional review board and the Ethics Committee of the American Psychological Association asking that they censure the Psychology Department for coercing unwilling subjects.

In response to the IRB's and the APA's inquiries, the Psychology Department countered with both pedagogical and pragmatic arguments. On pedagogy, the department argued that the experience of being a subject is quite important in helping introductory students understand the nature as well as the strengths and weaknesses of experimentation. Without a firsthand understanding of the process of experimentation, the department argued, it is difficult for students to evaluate the discussion and conclusions in their textbooks. On pragmatic research grounds, the department pointed out that the alternative of using paid subjects would result in distinctly nonrandom samples of the population. The use of nonrandom samples would introduce bias into the research results and would also destroy the comparability of such studies with previously conducted research. Further, the department argued, if the university research community had to rely solely on paid subjects, the progress of research would be slowed because fewer studies could be conducted. The efficient use of the university's resources committed to the advancement of knowledge would suffer as a result, as would the faculty's ability to bring fresh ideas into the classroom.

The department also pointed out that the student was not compelled to participate in the experiments, since she could have chosen to write a paper reviewing experimental methodology instead of participating in research. The department defended the practice of adding research hours when students miss appointments as a deterrent to minimize wasting both the experimenter's time and that of the other students.

Questions for Discussion

1. How should the university resolve this case? How should the APA handle it?

2. Of what relevance to your analysis is the availability of the paper-writing alternative?

3. Does the addition of an hour of experiments for missing an hour imply any inconsistency with a pedagogical aim?

4. Of what relevance to your analysis is the student debriefing? If you think that debriefing is an important aspect of the project design, what are the necessary elements of a debriefing that would make requiring participation in the experiments ethically unproblematic?

5. What if the student had already participated in one experiment and had found it to be a complete waste of time? What if her refusal came after having

participated in one or more experiments and having suffered an adverse reaction to the experiment?

6. Would it make a difference to you if Y had objected to both participating in the research and writing the alternative paper from the very beginning of the semester?

7. Suppose the student argued that she would serve only in experiments that were clearly for the good of society rather than for the good of the experimenters (e.g., studies designed to provide information that will ultimately help students or the mentally ill to function better). Does a situation involving "required participation" mandate such flexibility? Why or why not?

8. Under what circumstances, if any, can participation in research be required and yet be considered ethical? Do all cases in which participation is required involve coercion? What elements indicate coercion and which elements are necessary for avoiding coercion?

■ Case 2

X was a shy male undergraduate from a very conservative religious background. In his first semester at a small liberal arts college he took an introductory psychology class and became quite interested in personality and social psychology. The following semester he signed up for an advanced class called "Social Interaction as the Basis of the Self" taught by an experientially oriented female professor, Dr. Y. At the first class session Dr. Y informed the class that the course would function as a self-contained research project. They would be studying the social basis of the self by functioning as an encounter group centered on a topic of social interaction; the topic would be decided on by the class. All of the initial class sessions were to be videotaped and analyzed later in the semester in light of readings Dr. Y would assign on current theories of and research on social interaction and self. The students, as a class, would write the results of their research (i.e., their analysis of the videotapes) as a report at the end of the semester. Dr. Y then provided the students an opportunity to decline to participate in the class by withdrawing their enrollment. She explained that the topic of the course was likely to be personally sensitive and that because the students who chose to stay and participate would probably be revealing confidential information, withdrawal later in the course would not be allowed.

The discussion in the first session centered on selecting the topic of social interaction that the class would study. After considerable debate, the class picked the topic of stereotyped gender interactions as determinants of social selves. Dr. Y then asked each student to sign two forms. The first form asked for the student's consent to be videotaped. The second form asked the student to agree to certain conditions to enrollment in the course. Students were asked to agree (1) to participate in all sessions; (2) to hold private (i.e., not to communicate to any person outside the class) any and all information revealed in the class; and (3) to remain in the class for the duration of the semester (to ensure that private and possibly sensitive information would not be "leaked"). The form repeated Dr. Y's statement that students concerned about discussing personally sensitive information should opt out now.

At the beginning of the second class meeting, Dr. Y asked each student to recount stereotypes they had about the sexual behavior of the other gender. After the first several women participants spoke, X became distressed and attempted to leave the classroom. Dr. Y insisted that he stay, pointing out that he had agreed to participate for the duration of the semester and that leaving would constitute a breach of the agreement he had signed. Dr. Y further pointed out that X had participated in selecting the subject of gender interactions and that recognizing and breaking down stereotypes is a critical part of maturing, both as an individual and as a member of society.

X stayed and awkwardly took his turn in the session, but he did not return to the class. He subsequently filed a grievance with the university and with the APA arguing that he had been coerced against his will into listening to confidential aspects of others' lives and was forced to reveal inappropriate aspects of himself to strangers. He argued further that he had been coerced into staying in the class by a combination of social pressure and the fact that the class was the only one offered in psychology that semester that satisfied the university's distribution requirement. X complained that Dr. Y had no business experimenting with students' feelings, but should instead be teaching facts. Finally, X claimed that his relations with females in his church group had been irremediably harmed by his participation in the class, and that he was considering filing a lawsuit on these grounds.

In response, Dr. Y argued that the class rules had been clear from the beginning, that she had no control over the topics selected by the class, that X knew before signing the agreement that the class was going to study gender interactions and that the discussions were likely to contain very personal and private information, and that he was given the opportunity to decline to participate in the class during the first session. His agreement to the terms of the class constituted a waiver of his right to complain about it now, she claimed. Dr. Y argued that her role was to facilitate the group interaction and self-discovery possibilities in the class thereby producing the raw material for the academic analysis later in the course. She also pointed out that she did not provide academic content during the first part of the course because such knowledge would affect the group and individual processes they were trying to study.

Questions for Discussion

1. The federal regulations on human subjects require that participants in federally funded research must be free to withdraw from participation at any

time.[3] Should this concept be applicable to classroom-based research? Is the restriction on withdrawal coercive?

2. How would you resolve this issue as a member of the faculty grievance committee? As a member of the APA Ethics Committee?

3. For your analysis, does it matter what the gender balance of the class is? Does the gender of the student matter? What about the gender of the professor?

4. Suppose that Dr. Y points out that she is well known for teaching this course in this way and has done so every year for the past fifteen years, winning a teaching award in the process. Do these facts affect your analysis?

5. Is the signed agreement important to your analysis? What if X had not signed the agreement? What if there had not been an agreement, but that X had decided to drop the class after the date allowed by university rules?

6. What could Dr. Y have done differently to have made the class less vulnerable to ethical scrutiny?

■ Case 3

Dr. Z, a professor of clinical psychology at a medium-sized urban university, after receiving approval of the local IRB, ran an ad in the school newspaper seeking as paid subjects students who often felt depressed. After an extensive assessment battery and interview (for which all students were paid), 100 students were invited to sign up for a lengthy study that had to do with "the monitoring of depressive feelings."

Based on his review of the literature, Dr. Z had developed the hypothesis that a combination of regular physical activity and cognitive therapy involving the repetition of affirmations would decrease the frequency of depressive feelings. The study, which was double-masked,[4] contained four randomly assigned groups:

1. a control group;
2. a group engaging in physical activity;
3. a group receiving cognitive therapy; and
4. a group both engaging in physical activity and receiving cognitive therapy.

Each student signed a consent form committing the student to two months of three half-hour sessions per week plus a follow-up session. The student would receive $10 per hour of participation. Half of the money would be received at the end of the two months; the other half would be received after the six-month follow-up.

By the luck of the draw, three of the students were assigned to the control group. Control group members simply met and talked about their feelings. Over the course of several weeks, one of the students in the control group became increasingly more depressed and unhappy, eventually stopped showing up, and withdrew from school. She subsequently filed a grievance with the university and with the APA arguing that she had signed up for the experiment with the hope that it would help her, but that instead it had harmed her.

A second student in the control group, after talking with a psychology graduate student, discovered that he was not receiving the treatment that was hypothesized to be most effective and demanded that he be given that treatment immediately. The investigator refused, and this student also filed a grievance insisting that his rights were being abrogated and his mental health endangered by the psychologist's refusal to treat him.

A third student argued that she should be released from the study because it wasn't helping her and, further, that she had been coerced into participating in the experiment by a lack of money and the unrealized hope of getting better. She argued that she should be paid for her time to date because she had been paid in previous psychology experiments in which she had declined to continue when given the opportunity either to stop or continue.

Questions for Discussion

1. How would you evaluate each of the student's claims?
2. If each student's request or grievance is allowed, what are the implications for research related to therapy (e.g., clinical drug trials)?
3. Before consenting to participation in research, prospective subjects must be provided all information material to making a decision to participate. The specific elements of informed consent required by the federal regulations are provided in the Case Notes at the end of this chapter. What information should participants in this study be given as part of the informed consent process? What, if anything, was missing in this case?
4. Is the design of the study—random assignment to control and treatment groups—justified?
5. Should the students have been informed about the structure of the design (i.e., that participants would be assigned to one of the four groups and that each group would receive a different experience that might affect the number and extent of depressive feelings)? Should the students have been allowed to choose the group they wished to be in? Why or why not?
6. Could Dr. Z have taken steps to avoid these difficulties?

■ Case 4

Professor X, based on casual observations of himself and his family, found a perfect negative correlation between a person's ability to taste coal dust in the air and the ability to roll his or her tongue into a tube; two family members could taste coal dust but could not roll their tongues, while two family members showed the reverse combination. Over the next several days, Professor X informally collected data on the correlation using his colleagues, people who stopped into his office, and students in his laboratory. Inspired by the almost perfect correlation, he took data at the beginning of his two statistics classes. A colleague, after overhearing a summary report of the research in a hallway conversation, filed a grievance with the IRB and the APA on the grounds that the professor had not obtained approval for doing his research and, further, that he had used his status as a professor to coerce his students into participating.

In his defense, Professor X pointed out that when he started he was not certain there were any worthwhile data, and that if he had had to file a proposed experiment form with the IRB, he would never have taken the data. He also pointed out that he did not require any students to report the data to him; he simply inquired whether they could taste coal dust and roll their tongues. Further, he argued that the IRB wouldn't have approved the proposal anyway because he was not an expert in his field, had not done a complete literature survey, and would probably find no evidence in the literature for his hypothesis.

Questions for Discussion

1. What issues do you see here?

2. Do you think the professor acted inappropriately? Why or why not?

3. If you concluded that Professor X acted inappropriately, how, if at all, can scientists conduct informal inquiries?

4. Would it matter if Professor X had collected the data informally from friends and colleagues, but applied for IRB approval before asking students in his classes or laboratory?

5. Would it make any difference to your analysis if the data had hinted at a possible connection between these abilities and the early diagnosis of Huntington disease?[5] Would it matter if he had discovered the connection by asking people to provide him with information about hereditary diseases in their families?

■ Case 5

Based on her experience and the smattering of data in the literature, Professor A proposed the theory that black male criminals suffer from early childhood deprivation of attention by the male parent. She proposed to test her theory by comparing interviews of male ex-convicts with a group of men who had never been convicted of a crime matched on socioeconomic status, intact versus disrupted family, and presence versus absence of alternative male figures. To avoid basing her conclusions on potentially self-serving statements by the subjects, she also proposed to interview the parents and teachers of the subjects, many of whom she would have to track down.

A neighborhood group, Power for the People, filed a complaint with the university and the APA Ethics Committee claiming that the study was racist, coercive, designed to discredit the black community, harmful to the reputations of the families involved, and intended simply to advance Professor A's career without any concern for the people she might be damaging. They also argued that neither the university nor the APA Ethics Committee had the right to approve the research because neither committee's membership included any oppressed racial minorities (although both committees did include women and Asian Americans) and because the research took place in the black community. In their view, only the people from the black community had the right to approve the research, and not the university.

In her defense, Professor A argued before the IRB that her purpose was to help, not harm; her long term involvement with minority groups on campus had led her to this research, not hope for personal aggrandizement. If the data supported her hypothesis, she planned to apply for funding from the federal government to set up a program to help develop alternative male figures for families in the community. She also stated that she planned to speak with responsible community leaders about contacting family members, but that there was no single neighborhood or community that could be responsible for approving the research. She closed her remarks with the impassioned appeal, "If a perceived imbalance in power necessarily produces coercion and a victim, and these outcomes are viewed as more negative than any general long-term benefit that might accrue, the study and subsequent education of humans by any but novelists, artists, talk show hosts, and comics cannot go forward."

Questions for Discussion

1. Should the university allow this study to go forward as proposed?

2. Would it alter your conclusion if the university added one or more black community members to the IRB for the evaluation of this proposal, and required Professor A to form a community advisory group that would have a veto power over the study? Should all research studies have a specific advisory group drawn from the population being studied? (Note that the federal regulations allow IRBs to consult with non-IRB members when appropriate to ensure that subjects who are vulnerable are adequately protected. Consultants may not, however, vote on the approval of protocols. IRBs may add members who are knowledgeable about and experienced in working with particular populations as necessary to ensure that the interests of subjects are protected.)[6]

3. Is Professor A's race relevant to your analysis?

4. Would it make a difference to your analysis if the university and Profes-

sor A committed themselves to assisting the community to develop an appropriate intervention program if the results of the study appeared to warrant it?

5. Would you view the study differently if the subject of the study were different? For example:

a. What if the hypothesis concerned mentally ill blacks instead of ex-convicts?

b. What if the the hypothesis concerned blacks who had been successful in starting their own businesses?

c. What if the study concerned homosexuals instead of blacks?

6. Consider Professor A's concluding remarks to the IRB. Are they self-serving, or do they raise a significant issue?

Case Notes

Regulations pertaining to the protection of human participants in federally conducted or supported research have been adopted by 17 federal departments and agencies. The regulations require that subjects give their free and informed consent prior to participation and set out precisely what subjects must be told. Excerpts are given here.

In seeking informed consent, the following information shall be provided to each subject:

1. a statement that the study involves research, an explanation of the purposes of the research and the expected duration of the subject's participation, a description of the procedures to be followed, and identification of any procedures which are experimental;

2. a description of any reasonably foreseeable risks or discomforts to the subject;

3. a description of any benefits to the subject or to others which may reasonably be expected from the research;

4. a disclosure of appropriate alternative procedures or courses of treatment, if any, that might be advantageous to the subject;

5. a statement describing the extent, if any, to which confidentiality of records identifying the subject will be maintained;

6. for research involving more than minimal risk, an explanation as to whether any compensation and an explanation as to whether any medical treatments are available if injury occurs and, if so, what they consist of, or where further information may be obtained;

7. an explanation of whom to contact for answers to pertinent questions about the research and research subjects' rights, and whom to contact in the event of a research-related injury to the subject; and

8. a statement that participation is voluntary, refusal to participate will involve no penalty or loss of benefits to which the subject is otherwise entitled, and the subject may discontinue participation at any time without penalty or loss of benefits to which the subject is otherwise entitled.[7]

In addition, where appropriate, further information must be provided:

1. a statement that the particular treatment or procedure may involve risks to the subject . . . which are currently unforeseeable;
2. anticipated circumstances under which the subject's participation may be terminated by the investigator without regard to the subject's consent;
3. any additional costs to the subject that may result from participation in research;
4. the consequences of a subject's decision to withdraw from the research and procedures for orderly termination of participation by the subject;
5. a statement that significant new findings developed during the course of the research which may relate to the subject's willingness to continue participation will be provided to the subject; and
6. the approximate number of subjects involved in the study.[8]

IRBs can approve consent procedures that either do not include some of these elements or that alter some or all of them under certain limited circumstances. Primarily, the research cannot involve more than minimal risk to the subjects, the research must be such that it cannot be carried out unless the waiver or alteration is granted, and subjects must be provided with additional pertinent information after participation.[9] In essence, this exception allows IRBs to approve research that involves legitimate deceptive elements.

Notes

1. U.S. Department of Health and Human Services, "Protection of Human Subjects," *Code of Federal Regulations*, Title 45, Part 46.
2. See, e.g., Angoff 1985, Cohen 1982, Gamble 1982, Levine 1986: 80–82, Maloney 1984: chap. 7, Nolan 1989, and Shannon 1979.
3. 45 *Code of Federal Regulations* 46.116(a)(8).
4. A masked study is one in which either the investigator, the subject, or both do not know the identity of the treatment group to which the subject has been assigned. A double-masked study is one in which neither the investigator nor the subject knows the identity of the treatment group.
5. Huntington disease is a late onset genetic disease whose symptoms—severe mental and physical deterioration—do not appear until adulthood (usually after age 40). There is no effective treatment or cure for Huntington disease.
6. 45 *Code of Federal Regulations* 46.107.

7. Ibid., 46.116(a).
8. Ibid., 46.116(b).
9. Ibid., 46.116(d).

References

Angoff, Nancy R. "Against Special Protections for Medical Students." *IRB* 7(5): 9–10 (September/October 1985).

Cohen, Jeffrey M. "Extra Credit for Research Subjects." *IRB* 4(8): 10–11 (October 1982).

Freedman, Benjamin. "Equipoise and the Ethics of Clinical Research." *New England Journal of Medicine* 317(3): 141–145 (July 16, 1987).

Gamble, H. F. "Case Study: Students, Grades and Informed Consent." *IRB* 4(5): 7–10 (May 1982).

Levine, Robert J. *Ethics and Regulation of Clinical Research*. 2d ed. Baltimore: Urban and Schwarzenberg, 1986, pp. 80–82.

Maloney, Dennis M. *Protection of Human Research Subjects: A Practical Guide to Federal Laws and Regulations*. New York: Plenum Press, 1984, chap. 7.

Nolan, K. A. " 'Protecting' Medical Students from the Risks of Research." *IRB* 1(5): 9 (August/September 1989).

Shannon, Thomas A. "Case Study: Should Medical Students Be Research Subjects?" *IRB* 1(2): 4 (April 1979).

U.S. Department of Health and Human Services. "Protection of Human Subjects" *Code of Federal Regulations*, Title 45, Part 46.

U.S. Department of Health and Human Services. Public Health Service. National Institutes of Health. Office for Protection from Research Risks. *Protecting Human Research Subjects: Institutional Review Board Guidebook*. Washington, DC: Government Printing Office, 1993.

15 Behavior Control

EACH OF US controls the behavior of others. Sometimes the control is exerted by means of positive events such as food, money, or approval. At other times the control involves punishment via the application of aversive events, such as disapproval, social rejection, or hitting. With each of these forms of control, a controlling event can be either the application or withdrawal of a stimulus.

Behavioral control can often be innocuous or benign. We would all agree that there is nothing wrong with rewarding helpfulness by saying "thank you" or encouraging conversation by paying attention. However, behavior control may be more problematic when aversive stimuli are used or, as in behavior modification procedures, when the control is explicit and designed to change behavior in the direction that the controller considers more socially acceptable.

Behavior modification, especially if if involves aversive stimuli, may seem cruel or reminiscent of *A Clockwork Orange*, but such is not necessarily the case. When used properly, behavior modification procedures can often improve people's lives. Examples of successful applications are the use of aversive conditioning to prevent self-mutilation, the use of token economies in institutional environments (where residents' behavior may earn them tokens exchangeable for treats or privileges), and the use of disulfiram, a nausea-inducing drug, in the treatment of alcoholism.

The cases in this chapter raise ethical issues involved in behavioral modification, although only some of them involve research. All are concerned with the reduction or elimination of behavior that the controller considers undesirable. Two forms of behavioral control are used: either the administration of aversive stimuli (punishment) or the omission of positive reinforcers (e.g., "time out") if the undesirable behavior occurs.

The first three cases involve the use of disulfiram; the last four involve the use of aversive stimuli or positive reward omission in institutional settings. The cases vary in the severity of the behavior to be modified, the extent of the subject's informed consent, and the type of conditioning used. The final case focuses on the behavior being controlled and the nature of the control.

As you read through these cases, consider the ethical acceptability of the procedures used. Your conclusions may depend, among other things, upon your evaluation of individual versus societal rights and obligations, upon the identities of the person being controlled and of the controlling person or agency, upon the nature and purpose of the control, and, perhaps, also upon whether the controlling events are positive or aversive.

All persons subjected to behavioral control procedures lose some measure of personal freedom. Thus these cases all raise the question whether procedures that constrain personal liberty can be justified by the benefits they produce.

Background for Cases Involving Disulfiram

Disulfiram is a widely used drug that was introduced as a treatment for alcoholism in 1948.[1] Disulfiram produces an adverse drug-alcohol interaction, which discourages alcoholics who take it from consuming alcohol. It takes up to two weeks to metabolize. If alcohol is consumed during the two-week period when the disulfiram is in an active form in the individual's system, he or she will experience discomfort that can be extreme. The disulfiram-alcohol interaction first produces facial flushing and redness that spreads to the chest and limbs. This effect is followed by nausea, palpitations, and hypotension. Other side effects may include disorientation,[2] psychosis,[3] depression,[4] and a loss of libido.[5] The risk that the latter side effects will occur is small, but significant enough to require that disulfiram be administered only by prescription and pursuant to a physician's supervision.

The strength of the disulfiram treatment program rests in complete patient knowledge of the disulfiram-alcohol interaction. Patients will abstain from alcohol consumption in order to avoid the disulfiram-alcohol reaction. Disulfiram thus acts as a temporary self-imposed restraint on drinking. The disulfiram-induced period of abstinence allows the patient to receive other treatments, both psychological therapies and biochemical treatment such as antidepressant medications. Although when used in combination with other treatments a disulfiram program may be the most successful way to treat alcoholism, there is a debate in the scientific literature over the ultimate usefulness of disulfiram. While there are no statistics on effectiveness of this drug, experience has shown that it is ineffective unless combined with psychotherapy, and that its real effectiveness may rest solely on the patient's motivation. In addition, some clinics discourage the use of disulfiram because they fear patients might choose to remain on the drug, thus replacing alcohol dependency with disulfiram dependency.

■ Case 1

Julie is a business executive. At one time her alcohol consumption was limited to business lunches and occasional cocktail parties. As the demands of her job increased, however, so did her drinking. At first she felt there was no problem: she was just relieving stress. But soon she started having family problems and difficulty completing work. After being diagnosed as an alcoholic, Julie now feels she has a problem and seeks help, but she is unable to stay sober. Finding herself kicked out of her own home and in danger of losing her job, she consults with a number of medical doctors, including psychiatrists, seeking a cure for her alcoholism.

After considering a number of alternative treatments, Julie decides to try a disulfiram treatment program. She goes to a certified outpatient clinic where she is informed about the use of disulfiram, its side effects, and the treatment regimen. Treatment at the clinic includes the use of problem-solving and relapse prevention programs along with the disulfiram treatment.[6] Julie decides to become a patient.

Questions for Discussion

1. Is the use of disulfiram as a form of aversive behavioral conditioning ethically acceptable? Of what relevance is disulfiram's proven effectiveness?

2. As an alcoholic, is Julie capable of giving her informed consent to treatment for alcoholism? If not, who, ethically, can make the decision on her behalf?

3. Overall, is the treatment program Julie has entered ethical? Why or why not?

■ Case 2

Dr. Sober is interested in developing an effective disulfiram treatment program for alcoholism. He designs an experiment to test the effectiveness of disulfiram treatment in combination with various other social, psychological, and nonchemical treatments. He has four experimental groups. One receives the disulfiram treatment alone, and the others receive disulfiram in combination with one of three other types of treatment: alcohol education, relapse prevention, or Alcoholics Anonymous.

Subjects are volunteers who have sought treatment for alcoholism elsewhere, but who cannot afford to enter a clinic program. They have been referred to Dr. Sober by a number of different clinics and individual physicians. After being divided into the experimental groups, subjects are given a detailed explanation of the effects and usage of disulfiram. After the termination of the experiment, subjects have the option of receiving the most effective treatment (which might include disulfiram).

Questions for Discussion

1. What ethical responsibilities, if any, does Dr. Sober have to the subjects?

2. What ethical problems, if any, does the experiment present?

3. What issues of informed consent does this experiment raise?

4. The design of the experiment presents several questions: Does it matter how subjects are recruited into the study? Does it matter how subjects are assigned to the four alternative treatment groups? Is it acceptable for some of the people seeking treatment to be given possibly less effective treatments? What if the most effective treatment becomes evident before the end of the experiment? Must Dr. Sober then inform the other participants in the study of this development and allow them to participate in the more effective treatment strategy?

5. If you were a member of an IRB that was reviewing this protocol, would you vote to approve the research? (For a discussion of IRBs and their role in approving research, see the introduction to chapter 10 and the questions for discussion following the case, as well as the case notes for chapter 14.) What information, if any, would you want Dr. Sober to provide the IRB before making a decision? What changes, if any, would you request that Dr. Sober make?

■ Case 3

Felix is in court again. This is the second time he has appeared before a judge for drunk driving. This time, however, Felix is responsible for injuring someone. His car crashed into another and its driver had to be sent to the hospital. After a trial, which includes court-ordered physical and psychological examinations, the judge finds Felix guilty of drunk driving. Because the court-ordered examinations reveal that Felix is an alcoholic, Felix is given the option of mandatory treatment as an alternative to jail. Felix accepts the alternative and is ordered to undergo treatment for alcoholism with Dr. Jones. Dr. Jones is widely known for always treating his alcoholic patients with disulfiram.

Questions for Discussion

1. Is the judge's decision ethically justified? (Remember that Felix has not volunteered for this treatment, and that disulfiram's effectiveness depends upon the patient's motivation to be cured.) Does it matter that other courts have ordered different types of treatment for alcoholism than disulfiram? What if it were clear that disulfiram *is* the most effective treatment for alcoholism?

2. Has Felix given informed consent to disulfiram treatment in this case?

3. Is the use of aversive behavioral modification justified in this situation?

4. How are the three cases concerning disulfiram use alike and how are they different?

Background for Cases Involving Children

Behavior modification is also commonly used when working with "troubled" or "special" children. The procedures used can range from "time outs" (periods of time during which the child is isolated from all activities), to electric shock. It has been asserted that five considerations should govern the use of behavioral modification in working with children:

1. the necessity of stating goals in objective, measurable terms;
2. assurance that achieving the goal will benefit the child more than the school or institution;
3. a reasonable degree of certainty that achieving the goal is realistic for the student;
4. assurance that the program will help the student develop appropriate behaviors, not just suppress inappropriate ones; and
5. certainty that the behavior to be changed is not one that is protected by constitutional rights.[7]

■ Case 4

Lenny is a large teenage youth. Although he looks quite mature in stature, Lenny's mind has never caught up with his body. He is mentally retarded and lives in an institution. In addition to his mental handicap, Lenny has another severe problem: When he gets frustrated, he hurts himself. Because he is very strong, he can cause a lot of damage when he hits himself. The staff has exhausted almost every possible method to deter this behavior. One day Lenny hits himself with such force that he breaks his own jaw. The staff decides to try aversive behavioral modification, since it has been successfully used in cases similar to Lenny's. Lenny is fitted with a device that will allow trained and qualified members of the staff to deliver an electric shock to him when he exhibits self-mutilative behavior.

Questions for Discussion

1. Is the use of electric shock to alter Lenny's self-destructive behavior ethically appropriate? Why or why not? What if Lenny is physically harming other patients at the institution? If you think the treatment is ethical, what conditions or limitations, if any, would you place on its use?

2. From an ethical perspective, is there a difference between shock treatments that are administered by the staff and shocks either self-administered by the patient or set up to occur automatically when the patient exhibits certain behaviors?

3. In what respects is this case similar or dissimilar to the disulfiram cases?

4. What issues of informed consent does this case raise? How should informed consent be handled in this case?

■ Case 5

Rich is in grade school. He is like any other child, except that he cannot sit still. In class he gets up and walks around during lesson time. Surprisingly, Rich is still a good student. However, his behavior distracts the other children. His teacher, principal, and parents meet and agree that Rich must change his behavior. The school administration has recently been asked to participate in a study testing new techniques of behavior modification. The principal asks Rich's parents if they are willing to have Rich participate in the study, and they agree. At a meeting involving the study's researchers, the principal, teacher, and Rich's parents, a plan is adopted whereby the teacher will implement a procedure called "time outs." When Rich exhibits his restless behavior, the teacher will isolate him from the rest of the class. If this behavior continues, he will lose his recess privileges.

Questions for Discussion

1. Is the treatment proposed for Rich both appropriate and ethically acceptable? Why or why not? What if Rich's behavior continues for over a month? What if Rich is physically harming other students?
2. How does this kind of behavioral modification differ from that used with Lenny? Is it more or less acceptable? Why?
3. What issues of informed consent does this case raise? How should informed consent be handled in this case?

■ Case 6

Michelle is a junior at a high school located in a small conservative town. She is a good student and enjoys extracurricular activities of all kinds. In her high school, students are required to wear uniforms. In addition, they cannot wear jewelry and must comply with established grooming standards. Michelle feels that her high school is too authoritative and decides to have half her head shaved and the other half dyed plaid. Michelle is now attracting much attention. The school administrators are alarmed at Michelle's behavior; they conclude that she is a rebel stirring up trouble. They decide to use a form of "time out" in an effort to change her behavior. They warn her that if her behavior and appearance do not change, she will be barred from all extracurricular activities. If this does not work, she will have to spend her study hall and lunch hours in the main office.

Questions for Discussion

1. Is the treatment of Michelle ethically justified? Why or why not? What if the student were male and the objection was to his hair length? What if Michelle's appearance was required by her religious beliefs?

2. How should informed consent be dealt with in this case? Would the case be different if the administrators had met with Michelle's parents and they had given their consent? Why or why not?

3. How is Michelle's case different from Rich's or Lenny's case? Are these ethically relevant differences?

■ Case 7

Gary Travis is a psychiatrist at Mammoth State Mental Hospital. He has a patient named Alfred Tunis, who has been diagnosed as personality disordered, and who has been at the hospital for two years. Alfred has recently begun disrupting any ward he is placed on; when people come near him he frequently and unpredictably attacks them with any object that is handy (pencils, pens, utensils from the dining hall, lamps). He has severely injured another patient who did not defend herself, and has stabbed both a nurse and Dr. Travis in the arm. Dr. Travis has several choices he can make in terms of treatment:

a. increase Alfred's dosage of tranquilizing drugs;

b. isolate Alfred from the other patients until he calms down;

c. use a personal shocker to keep Alfred at a distance when he makes a threatening move toward staff;

d. give all the patients a shocker to use on Alfred;

e. increase Alfred's time outside, an activity Alfred very much enjoys, contingent on his not attacking people;

f. take away Alfred's access to utensils;

g. remove privileges—such as watching television or playing basketball—when Alfred attacks someone;

h. meet with the patients and organize them into a group to deal with Alfred when he threatens any of them;

i. give Alfred ice water baths whenever he attacks anyone;

j. increase the number of hours Alfred gets psychotherapy in hopes that the reason for his attacks can be found;

k. put Alfred in a back ward with the dangerously ill;

l. notify Alfred's guardian that he cannot be retained at the hospital because of his violence; or

m. sue Alfred and Alfred's guardian for personal damages and mental suffering.

Questions for Discussion

1. What ethical considerations are involved in choosing or not choosing each of these treatments?

2. Notice that each of these solutions has multiple effects, both in decreasing Alfred's access to some events or states and increasing his access to others. How do you evaluate these multiple effects?

3. What if Alfred's disruption consisted of singing Beatles songs loudly?

4. What if Alfred were your brother?

5. What if Alfred were a woman?

6. What if Alfred were a child in your classroom?

7. What if Alfred were your next-door neighbor?

Notes

1. A description of alternative forms of treatment for alcoholism is contained in Mendelson and Mello 1987.

2. Disorientation might result as to place or time. The person may lose track of time and not appear to be "with it." The patient may be confused at times, showing some evidence of slowed information processing, for up to one hour. This disorientation effect is rare.

3. This potential side effect is extremely rare and would occur only in an individual who has either a predisposition toward, or a history of, psychotic disorder. A period of psychosis is marked by a significant break from reality, where the individual displays a thought disorder (e.g., a delusion or complete incoherence) and/or hallucinations. This effect can be very severe and requires immediate medical attention and medication.

4. Depression may result from some of the pharmacological effects of disulfiram, but may also be due to the fact that once the individual cannot have access to alcohol in order to ameliorate anxiety or depression, an already existing depressed mood may increase. (In many cases, depression occurs with alcoholism.) When depression occurs, it is usually of a moderate severity and requires psychological or psychiatric treatment.

5. The individual may be less interested in sexual activity. This side effect may last for the entire period of disulfiram treatment.

6. Alcoholism treatment programs presently lack uniformity. Approaches include general treatment programs, such as Alcoholics Anonymous, Twelve Step, and humanistic therapies focused on problem solving, as well as more specific programs such as cue-exposure desensitization treatment and the relapse prevention treatment mentioned in the text.

7. Alberto and Troutman 1986: 47.

References

Alberto, P. A., and Troutman, A. C. *Applied Behavior Analysis for Teachers*. Columbus, OH: Merrill, 1986.

Mendelson, Jack, and Mello, Nancy, eds. *The Diagnosis and Treatment of Alcoholism*. New York: McGraw-Hill, 1987.

PART IV

Cases in the Humanities: History

16 The Historian's Code of Ethics

THE *New York Times* recently stated that "in academia, the use of another person's words or ideas without attribution is one of the most serious offenses."[1] This chapter explores the ethics of plagiarism in academia from the perspective of professional codes of ethics. Many professions have a code of conduct that guides practitioners. The introduction to the American Historical Association's *Statement on Standards of Professional Conduct* states that despite the diversity of the historical profession, "all historians should be guided by the same principles of conduct" (American Historical Association 1993: 1).[2] The major objective of this chapter is to examine the purposes and practical implications of the "rules and regulations" set out by the members of the AHA with respect to plagiarism.

The term *plagiarism* refers to a number of unethical practices. The principal ones are appropriating the work of another and claiming it as one's own; quoting material from a source without using quotation marks or citing the author(s); and using the words or ideas of another author without providing a citation or proper attribution. Plagiarism does not always involve clear-cut cases of one person's use of another's words or ideas without attribution. It ranges from deliberate appropriation of the work of another (stealing) to the inadvertent failure to cite properly the work of another. In addition, plagiarism is sometimes difficult to differentiate from situations in which more than one person independently arrives at the same or similar ideas.

Like the forms of plagiarism, the sanctions or penalties for plagiarism vary from case to case. Sanctions against students range from reprimands to expulsion. Similarly, professionals who have been found to have plagiarized may lose their university positions as well as their reputations as honest scholars.

As you read the case vignettes and the statement on plagiarism of the AHA, think about:

1. the ethical issues involved in plagiarism;
2. how you think plagiarism should be dealt with (both at the student and faculty level);
3. whether the AHA statement adequately addresses the ethical issues involved in plagiarism; and
4. the reasons for your position.

■ Case 1

The following text is excerpted from the American Historical Association's "Statement on Plagiarism and Related Misuses of the Work of Other Authors" (American Historical Association 1993: 13–16).[3]

I. Identifying Plagiarism and Other Misuses

The word *plagiarism* derives from its Latin roots: *plagiarius*, an abductor, and *plagiare*, to steal. The expropriation of another author's text, and the presentation of it as one's own, constitutes plagiarism, and is a serious violation of the ethics of scholarship. It undermines the credibility of historical inquiry.

In addition to the harm that plagiarism does to the pursuit of truth, it is also an offense against the literary rights of the original author and the property rights of the copyright owner. Detection can therefore result not only in academic sanctions (such as dismissal from a graduate program, termination of a faculty contract, denial of promotion or tenure) but also civil or criminal prosecution. As a practical matter, plagiarism between scholars rarely gets into court. Publishers are eager to avoid adverse publicity, and an injured scholar is unlikely to seek material compensation for misappropriation of what he or she gave gladly to the world. The real penalty for plagiarism is the abhorrence of the community of scholars.

The *misuse* of the writings of another author, even when one does not borrow the exact wording, can be as unfair, as unethical, and as unprofessional as plagiarism. Such misuse includes the limited borrowing, without attribution, of another historian's distinctive and significant research findings, hypotheses, theories, rhetorical strategies, or interpretations, or an extended borrowing even with attribution. Of course, historical knowledge is cumulative, and thus in some contexts, such as textbooks, encyclopedia articles, or broad syntheses, the form of attribution, and the permissible extent of dependence on prior scholarship, will be different than in more limited monographs. As knowledge is disseminated to a wide public, it loses some of its personal reference. What belongs to whom becomes less distinct. But even in textbooks a historian should acknowledge the sources of recent or distinctive findings and interpretations, those not yet a part of the common understanding of the profession, and should never simply borrow and rephrase the findings of other scholars.

Both plagiarism and the misuse of the findings and interpretations of other scholars take many forms. The clearest abuse is the use of another's language without quotation marks and citation. More subtle abuses include the appropriation of concepts, data, or notes all disguised in newly crafted sentences, or reference to a borrowed work in an early note and then extensive further use without attribution. All such tactics reflect an unworthy disregard for the contributions of others.

II. Resisting Plagiarism

All who participate in the community of inquiry, as amateurs or as professionals, as students or as established historians, have an obligation to oppose deception. This obligation bears with special weight on the directors of graduate seminars. They are critical in shaping a young historian's perception of the ethics of scholarship. It is therefore incumbent on graduate teachers to seek opportunities for making the seminar also a workshop in scholarly integrity. After leaving graduate school, every historian will have to depend primarily on vigilant self-criticism. Throughout our lives none of us can cease to question the claims our work makes and the sort of credit it grants to others.

But just as important as the self-criticism that guards us from self-deception is the formation of work habits that protect a scholar from plagiarism or misuse. The plagiarist's standard defense—that he or she was misled by hastily taken and imperfect notes—is plausible only in the context of a wider tolerance of shoddy work. A basic rule of good notetaking requires every researcher to distinguish scrupulously between exact quotation and a paraphrase. A basic rule of good writing warns us against following our own paraphrased notes slavishly. When a historian simply links one paraphrase to the next, even if the sources are cited, a kind of structural misuse takes place; the writer is implicitly claiming a shaping intelligence that actually belonged to the sources. Faced with charges of failing to acknowledge dependence on certain sources, a historian usually pleads that the lapse was inadvertent. This will be easily disposed of if scholars take seriously the injunction to check their manuscripts against the underlying texts prior to publication. Historians have a right to expect of one another a standard of workmanship that deprives plagiarism or misuses of their usual extenuations.

The second line of defense against plagiarism is organized and punitive. Every institution that includes or represents a body of scholars has an obligation to establish procedures designed to clarify and uphold their ethical standards. Every institution that employs historians bears an especially critical responsibility to maintain the integrity and reputation of its staff. This applies to government agencies, corporations, publishing firms, and public service organizations like museums and archives, as surely as it does to educational facilities. Usually, it is the employing institution that is expected to investigate charges of plagiarism or misuse promptly and impartially and to invoke appropriate sanctions when the charges are sustained.

Many learned professions are just beginning to think seriously about the need for general policies on fraudulent research and writing. Usually, employing institutions tend to respond to each case in an ad hoc manner, with responses ranging from extreme indulgence to uncompromising severity. Penalties for scholarly misconduct should vary according to the seriousness of the offense. A persistent pattern of deception justifies termination of an academic career; some scattered misappropriations may warrant only a public disclosure. What is troubling is not the variation in responses but rather the reluctance of many scholars to speak out about the possible offenses that come to their notice. No one advocates hasty or ill-founded accusations, and the protections of due process should always apply. If, however, charges of plagiarism or gross misuse are sustained by an investigating committee, its findings should ordinarily be made public. When appraising manuscripts for publication, reviewing books, or evaluating peers for placement, promotion, and tenure, the trustworthiness of the historian should never be overlooked. After all, scholarship flourishes in an atmosphere of openness and candor, which should, in our opinion, include the scrutiny and discussion of academic deception.

Questions for Discussion

1. What are the ethical issues presented in the AHA statement?

2. Why is plagiarism an ethical issue?

3. What distinctions is the AHA drawing between "plagiarism" and "other misuses"? Are these distinctions valid?

4. What are the likely results of violating the guidelines set forth in the statement?

5. How well do you think these guidelines protect against the abuses with which they were designed to deal?

6. More generally, what should be the purpose of a professional code of ethics?

7. What sanctions are appropriate for plagiarism and other misuses of intellectual property? Listed below are some possibilities. Discuss the appropriateness or inappropriateness of each, justifying your conclusions and suggesting alternatives where possible.

a. No penalty should be imposed, at least for a first offense.

b. The offender should be ostracized by the community of scholars of which he or she has been a part.

c. The offender should be removed from the community of scholars of which he or she is a part.

d. Civil and criminal sanctions should be imposed on the offender.

e. Student offenders should fail the course in which they plagiarized.

f. Graduate student offenders who commit plagiarism in writing their theses should be prevented from obtaining their degrees.

g. Faculty offenders should be denied tenure.

h. Faculty offenders should be threatened with dismissal.
i. Some combination of the penalties listed above should be imposed.
j. Some alternative to these penalties should be imposed.

8. Should sanctions differ depending on whether the offender is a student or a faculty member? Why?

9. Should sanctions for plagiarism differ depending on the seriousness of the offense?

10. What should be the role of a student's mentor (professor, graduate instructor, advisor, etc.) in guarding against plagiarism?

11. As the AHA decision to distinguish between "plagiarism" and "other misuses" indicates, questions of the proper treatment of intellectual property are complex. Given this complexity, how can these kinds of ethical breaches be regulated most effectively?

12. Do you agree with the statement's authors that students "have an obligation to oppose deception"?

a. Why or why not?
b. If you disagree, how would you amend the statement to reflect the obligations you consider appropriate?
c. If you agree, to what lengths should this obligation extend?
i. Is it enough to notify the professor if you suspect that a classmate has plagiarized on a paper? Why?
ii. Should you also be willing to testify or submit a written statement in connection with a proceeding that might be brought against the suspect? Why?
iii. Should you also be obligated to participate in such proceedings? Why?
iv. What are students' obligations with respect to suspected misconduct on the part of a faculty member? Are these obligations realistic?

■ Case 2

Mary is an assistant professor of modern Japanese economic history. Everyone in the field acknowledges her to be one of the brightest young minds in the profession. Her first book, based on her dissertation, is a study of post–World War II economic recovery. Following the customary practice, before agreeing to publish Mary's book the publisher sent the manuscript to several experts in Mary's field for them to review and provide critical comments. The referees all agreed that it is a first-rate, pathbreaking work. The press therefore accepted the book for publication.

Bill, Mary's counterpart at another university, has eagerly begun to read Mary's book, which has now been published. As he reads, he gets the feeling that something is not right. He is sure that he recognizes some of the text. Certain that it seems familiar because Mary has published articles and delivered papers at conferences based on chapters of the book, Bill decides to let it go. But after finishing the book, he is left with a gnawing feeling that the familiar passages are not Mary's own words. He decides to look at her published work and satisfies himself that her articles are not the source of the familiar text.

A year has gone by. During that year, Mary's book has received glowing reviews, both in professional journals and in the popular press. Her new research has attracted a good deal of attention from her colleagues at other universities across the country. In the meantime, while doing some of his own research work, Bill looks at his notes from an unpublished dissertation and finds a quote that sounds familiar from another context. On reflection, he realizes what that source was: Mary's book. He pulls the book off the shelf, riffles through, and, sure enough, there it is. Mary appears to have taken portions of the dissertation and used them as if she had written them herself. There are no quotation marks around the text in question, nor is there a cite to this or any other source. He wonders what other portions of her book have been taken from other sources and also about the integrity of the articles she has published. Bill is in a quandary: What should he do?

1. According to the AHA guidelines, has Mary committed plagiarism?

2. What should Bill do about his discovery (e.g., confront Mary directly, contact her department chair, contact the AHA, contact the author[s] whose work he believes Mary has plagiarized, do nothing)? Explain.

3. If Mary is found to have plagiarized other scholars' work, what professional consequences should result?

4. What responsibility should Mary's dissertation supervisor have for her work?

Suppose that Bill brings his suspicions to Mary's department chair. The chair talks the matter over with Mary. She explains that she did not intend to plagiarize; that her note taking must not have been as careful as it should have been.

Should Mary's explanation end the matter? What further obligation does either the institution or Bill have?

Suppose that the chair is not satisfied with Mary's explanations. She decides to initiate an investigation into Bill's allegations. The investigation results in a finding that Mary not only plagiarized this source and several others in her book, but also numerous other sources in her other published works. Technically, Mary should be asked to leave the university. But everyone agrees that despite these flaws, her work is brilliant. The university committee, together with the department chair, decides to allow Mary to stay on, but to require that her work be reviewed by a more senior scholar in the department, so that she can learn better research methods.

1. Is the outcome of this case appropriate? On what do you base your answer?

2. If the outcome of this case is not appropriate, what would be an appropriate result?

■ Case 3

Alfred is an undergraduate student who has a paper due tomorrow in his history course. He has been unable to write it because he has had the flu. He did the reading for the paper, but got sick before he had a chance to write it up. The professor will not accept late work for any reason; instead, students are allowed to drop one grade. Alfred was planning to drop the C he got on the first midterm exam: He is a pre-med, and, he believes, won't get into medical school unless he gets an A in the class. Alfred feels that the C was not a good measure of his performance; he's worked very hard in this course. Distraught, Alfred phones his friend Bella, who took the same course last year and wrote a paper on the same book Alfred was going to write on. Bella suggests that Alfred take her paper and just "rework" it a little.

Questions for Discussion

1. What should Alfred do?

2. If Alfred takes Bella's paper and hands it in without any changes except the name, is he behaving unethically? If so,

a. What has he done wrong;
b. Why is it wrong; and
c. What should the consequences for Alfred's wrongdoing be? What, for example, would you do if you were the professor?
   i. Would you do nothing? If so, why?
   ii. Would you give Alfred an "F" on the paper? Why or why not?
   iii. Would you fail Alfred in the course? Why or why not?
   iv. Would you report Alfred to the university administration or campus ethics committee? Why or why not?

3. If Alfred takes Bella's paper and hands it in with some minor changes, is he behaving unethically? If so,

a. What has he done wrong?
b. Why is it wrong?

c. What should the consequences for Alfred's wrongdoing be? What, for example, would you do if you were the professor?

i. Would you do nothing? If so, why?
ii. Would you give Alfred an "F" on the paper? Why or why not?
iii. Would you fail Alfred in the course? Why or why not?
iv. Would you report Alfred to the university administration or campus ethics committee? Why or why not?

4. What use can Alfred ethically make of Bella's paper?
5. Would your answer to Questions 1, 2, and 3 change if the reason Alfred didn't write the paper was that he had three exams the same week his paper was due? Why or why not?
6. Would your answer to Questions 1, 2, and 3 change if the reason Alfred didn't write the paper was that his grandmother had died, and he had to go home for the funeral? Why or why not?
7. Would your answer to Questions 1, 2, and 3 change if Alfred was a biology major and just couldn't "get" history? Why or why not?

■ Case 4

Catherine, a graduate student in history, is writing her dissertation. She has completed the research for the thesis and has her arguments well laid out. Both she and her advisers believe her work constitutes a new approach to the subject. She has completed the first two chapters of what will be a five-chapter dissertation.

As she begins to write chapter three, Catherine comes across an unpublished dissertation that deals with her subject that, despite diligent research, she had not found before. To her dismay, she finds that it expresses a number of the same ideas she thought were her own.

Catherine is convinced that her advisers would never have heard of or seen the thesis since she came upon it serendipitously, and as far as she can determine, it has never been cited anywhere. "I arrived at these ideas independently," she thinks. "Why shouldn't I get credit for them? Besides, acknowledging the other thesis would mean completely reworking my dissertation, and I'm so close to finishing it! Everyone agrees that it's good, and this guy's work will never be published."

Questions for Discussion

1. What should Catherine do about her discovery?

a. Should she act as though the unpublished dissertation does not exist?
b. Should she use the unpublished dissertation, but without acknowledging that it reflects the same ideas as her own work?
c. Should she acknowledge the dissertation properly and adjust her own

work to accommodate the ideas in the unpublished dissertation, even if it means largely starting over?

d. What else might she do?

2. What reasons support your answer to Question 1?

3. Suppose that as Catherine is completing her dissertation, a new book on the same subject is published. What should Catherine do?

4. What ethical issues does this case raise?

Notes

1. *New York Times* (July 3, 1991), p. A8.

2. In addition to the statement from which it derives its title, this booklet contains all of the guidelines on professional conduct and practice developed by the AHA's Professional Division. Included are the association's procedures for enforcing the association's standards of professional conduct and the following documents: "Statement on Plagiarism and Related Misuses of the Work of Other Authors," "Advisory Opinion Regarding the Harassment of Job Candidates," "Statement on Interviewing for Historical Documentation," "Statement on Discrimination and Harassment in Academia," "Advisory Opinion Regarding Conflict of Interest," and "Statement on Diversity in History Teaching."

3. Reprinted with permission. Copies of this and other AHA statements are available from the American Historical Association, 400 "A" St., S.E., Washington, DC 20003-2422.

Reference

American Historical Association. *Statement on Standards of Professional Conduct.* Washington, DC: American Historical Association, 1993.

17 The Use and Interpretation of Historical Documents

THE EXERCISE IN this chapter is designed to explore ethical problems associated with the use of historical documents. Our working assumption is that accurate use of sources is a key moral principle for historians. Following is part of a speech delivered by William Wilberforce, M.P., in the British House of Commons on May 12, 1789.[1] After you have read it carefully, you will be asked to read a student essay based on it and to assess the accuracy of the student author's claims.

The Text of Wilberforce's Speech

1 I must speak of the transit of the slaves to the West Indies. This I
2 confess, in my opinion, is the most wretched part of the whole subject. So
3 much misery condensed in so little room is more than the human imagination
4 had ever before conceived. . . . Let anyone imagine to himself six or seven
5 hundred of these wretches, chained two and two, surrounded with every object
6 that is nauseous and disgusting, diseased, and struggling under every kind of
7 wretchedness! How can we bear to think of such a scene as this? One would
8 think it had been determined to heap upon them all the varieties of bodily pain
9 for the purpose of blunting the feelings of the mind; and yet in this very
10 point—to show the power of human prejudice—the situation of the slaves has
11 been described by Mr. Norris, one of the Liverpool delegates, in a manner
12 which, I am sure, will convince the House how interest can draw a film over
13 the eyes so thick that total blindness could no more. . . . 'Their apartments,'
14 says Mr. Norris, 'are fitted up as much for their advantage as circumstances
15 will admit. The right ankle of one, indeed, is connected with the left ankle of
16 another by a small iron fetter, and, if they are turbulent, by another on their
17 wrists. They have several meals a day—some of their own country provisions
18 with the best sauces of African cookery—and, by way of variety, another meal
19 of pulse etc. according to European taste. After breakfast they have water to
20 wash themselves, while their apartments are perfumed with frankincense and
21 lime-juice. Before dinner they are amused after the manner of their country.
22 The song and dance are promoted,' and, as if the whole was really a scene of
23 pleasure and dissipation, it is added that games of chance are furnished. 'The
24 men play and sing, while the women and girls make fanciful ornaments with
25 beads which they are plentifully supplied with.'
26 Such is the sort of strain in which the Liverpool delegates, and
27 particularly Mr. Norris, gave evidence before the Privy Council. What will
28 the House think when by the concurring testimony of other witnesses the true

29 history is laid open. The slaves, who are sometimes described as rejoicing at
30 their captivity, are so wrung with misery at leaving their country that it is the
31 constant practice to set sail in the night, lest they should be sensible of their
32 departure. The pulse, which Mr. Norris talks of, is horse-beans; and the
33 scantiness both of water and provision was suggested by the very Legislature
34 of Jamaica in the report of their committee to be a subject that called for the
35 interference of Parliament. Mr. Norris talks of frankincense and lime-juice,
36 when the surgeons tell you the slaves are stowed so close that there is not
37 room to tread among them and when you have it in evidence . . . that even in a
38 ship which wanted two hundred of her complement the stench is intolerable.
39 The song and dance, says Mr. Norris, are promoted. It had been more fair,
40 perhaps, if he had explained that word 'promoted.' The truth is that, for the
41 sake of exercise, these miserable creatures, loaded with chains, oppressed with
42 disease and wretchedness, are forced to dance by the terror of the lash and
43 sometimes by the actual use of it. 'I,' says one of the other evidences, 'was
44 employed to dance the men, while another person danced the women.' Such,
45 then, is the meaning of the word 'promoted.' And it may be observed, too,
46 with respect to food, that an instrument is sometimes carried . . . in order to
47 force them to eat. . . . As to their singing, what shall we say when we are told
48 that their songs are songs of lamentation upon their departure. . . .
49 In order, however, not to trust too much to any sort of description, I
50 will call the attention of the House to one species of evidence which is
51 absolutely infallible. Death, at least, is a sure ground of evidence; and the
52 proportion of deaths will not only confirm, but, if possible, will even
53 aggravate our suspicion of their misery in the transit. . . . Upon the whole . . . here
54 is a mortality of about fifty percent and this among negroes who are not
55 bought unless quite healthy at first, unless—as the phrase is with cattle—they
56 are 'sound in wind and limb.' How, then, can the House refuse its belief to
57 the multiplied testimonies before the Privy Council of the savage treatment of
58 the negroes in the Middle Passage? Nay, indeed, what needs is there of
59 any evidence? The number of deaths speaks for itself and makes all such
60 inquiry superfluous.

The Student Essay

Assume that the following is part of a student essay on the African slave trade. It is based, in one way or another, upon the quotation provided above. It is your task to read each sentence and judge whether each sentence uses the material properly. If you decide that the text was not used properly, explain why.

1 Mr. Wilberforce and Mr. Norris held differing views on the African slave
2 trade.[1] William Wilberforce did not object to the slave trade per se, but he believed
3 that the conditions of transit were a problem, "the most wretched part of the whole
4 subject." Wilberforce reported that it was the constant practice of slave-traders to set
5 sail "in the night lest they [the slaves] should be sensible of their departure." In
6 contrast, Mr. Norris, a Liverpool trader, said that the slaves are not fettered at the
7 wrist, even "if they are turbulent." Wilberforce accused Norris of lying to the Privy
8 Council about the conditions of slave transportation because he did not want any
9 government interference in the trade.
10 The slave traders tried to take good care of their human cargo: they
11 provided them with good food, exercise, and entertainment on board the ships
12 that carried them to the West Indies. Wilberforce said that the slave quarters
13 on board ship were scented with "frankincense and lime-juice," and their food
14 included "the best sauces of African cookery." In order to keep the slaves
15 healthy, the slave traders hired men who whipped the slaves to force them to
16 exercise.[2] The slaves did not want to exercise and were forced to do so
17 by their owners.[3] Wilberforce's racism can be seen in the way he likened
18 African slaves to cattle, "sound in wind and limb."
19 Healthy Africans, after spending some time on slave ships in transit to
20 the New World, came down with diseases that often killed them.[4] The death
21 rate for slaves on the trip from Africa to the New World was 50%. Recent
22 historians have found that the mortality of Africans being transported to the
23 West Indies varied from 10% to 22%.[5] Wilberforce inflated the mortality
24 statistics in order to strengthen his case against the slave trade. Recently,
25 historians have argued that the mortality rates on slave ships were worse for
26 white seamen than they were for the slaves being transported.[6] Had
27 Wilberforce known this it would have destroyed his arguments against the
28 slave trade, because he thought "Death . . . is a sure ground of evidence."

1. William Wilberforce, "Speech before the House of Commons," *Parliamentary History* 28: 45–47, as quoted in *Cardinal Documents in British History*, R. L. Schuyler and C. C. Weston, comps. (Princeton, NJ: D. Van Nostrand, 1961), pp. 153–155.
2. Ibid.
3. Ibid.
4. Ibid.
5. Philip D. Curtin, *The Atlantic Slave Trade: A Census* (Madison: University of Wisconsin Press, 1969), p. 279.
6. Ibid., pp. 279, 282–283.

Note

1. William Wilberforce, "Speech before the House of Commons," *Parliamentary History* 28: 45–47, as quoted in Schuyler and Weston 1961: 153–155.

References

Curtin, Philip D. *The Atlantic Slave Trade. A Census*. Madison: University of Wisconsin Press, 1969, p. 279.

Wilberforce, William. "Speech before the House of Commons," *Parliamentary History* 28: 45–47, as quoted in *Cardinal Documents in British History*, R. L. Schuyler and C. C. Weston, comps. Princeton, NJ: D. Van Nostrand, 1961, pp. 153–155.

18 Oral Historians Meet the Media

ARE THERE SPECIAL or additional ethical demands placed on the work of an oral historian that are not applicable to the rest of the history profession? Michael Frisch and Dorothy Watts suggest that there are, in part because oral historians, unlike others, are involved in generating as well as interpreting texts. The "essential dialogical nature" of oral history, they allege, requires oral historians to be aware of and (to some extent) take responsibility for the way the statements of their subjects are used (Frisch and Watts 1980: 88–110, 101).

This case explores what happened when a group of oral historians contracted with the *New York Times* to write a story about the effects of widespread unemployment in Buffalo, New York. The journal excerpt that follows this introduction was written by the oral historians involved in the project after the *Times* published their research article. The authors argue that the story as published fails to reflect faithfully the sense and meaning of their subjects' statements.

All historians have a duty to represent accurately the data contained in the sources they use in their articles. For oral historians, in particular, living subjects are their sources, and taped conversations with those subjects constitute their data. The authors of the Buffalo study point out that

> oral historical material is produced in an interview situation, one in which the subject is triangulated between the interviewer and the experience being discussed. No matter how controlled the schedule of questions, the information is produced in a dialogue between individuals, each with a social position and identity, engaging in a conversation that exists at a necessary remove, in time or social space, from the experience being discussed. This is, of course, a fundamentally different relationship than usually exists between historians and the mute and frozen documents of the past; it has generally been discussed as an obstacle to objectivity surmountable through care and precision (Frisch and Watts 1980: 90).

What special duties do oral historians have, if any, to their living subjects to insure that their thoughts and ideas are reflected accurately in published form? Can historians fulfill their responsibilities to reflect accurately the information in their sources if they do not have final control over published content? In other words, is it ethical for a historian to agree to an arrangement in which he or she relinquishes control over the content of the publication? What special professional responsibilities, if any, do oral historians have to protect the integrity of their work against popularization by the media?

The American Historical Association's "Statement on Standards of Professional Conduct" states that "historians should . . . advise their readers of the conditions and rules that govern their work. They also have the obligation to decline to make their services available when policies are unnecessarily restrictive" (American Historical Association 1993: 2). As you read the following case, consider whether the oral historians involved in the Buffalo unemployment study heeded these guidelines.[1]

Oral Historians Meet the Media

In 1974–75, SUNY-Buffalo's American Studies Program was commissioned by the *New York Times Magazine* to prepare an oral documentary, in the manner of Studs Terkel, on the subject of unemployment in Buffalo, New York. This was eventually published as the magazine's cover story on February 9, 1975. What we will relate in some detail is a series of conflicts that developed between our group and the *Times* in the final stage of editing, conflicts about the basic content and composition of the article.

* * *

The early stages of the story can be briefly related. The *Times* commissioned the article at a time when the rising unemployment rate was just beginning to be recognized as a front-page problem. The magazine editors wanted a Terkelesque "bottom-up" perspective, using Buffalo—then "boasting" the highest unemployment rate in New York, and one of the highest in the nation—as a case study. When Terkel himself declined, our group was invited to design and submit an article.

From the start, we were under severe deadline pressure, which permitted us the dubious but real luxury of having no time to worry much about method or technique. We assembled a large group of faculty and students, improvised some quick training sessions, fanned out to conduct as many interviews as we could, developed ways of mapping our progress, redirecting our efforts as needed, and moved from tapes to transcripts to various proposed edits, going back and forth between the large group and a smaller editorial committee. Finally, we produced a documentary article which we felt reflected accurately what we had heard in the more than eighty interviews conducted, and in the many hundreds of pages of transcripts they generated.

The evolution of this process could be an instructive article in its own right, but for the topic at hand a number of points about it are particularly important. In the preparation process, we did not focus at all on consciousness as such. Rather, we sought to cast as wide a net as we could in every dimension, and see what we learned about people's experience of the opinions about unemployment. Our approach was, however, guided by a number of critical concerns and assumptions. For one, we were anxious not to be trapped by a literal, narrow definition of unemployment and whose experience was relevant to it. Thus, we made sure to interview men and women of all ages and a variety of backgrounds, not only the male factory workers of conventional imagery. And we tried to gauge the impact of unemployment as a more general crisis in work, family life, and society. We therefore interviewed not only those actually laid off regular jobs, but also the underemployed and the unemployable, whether too old, too young, too unskilled, or too overskilled for existing labor markets. We interviewed as well family members affected by the unemployment of others, officials and civil servants dealing with the problem, and so on.

A second major concern involved the ultimate public presentation of the piece; we were worried that the article's local focus might overly particularize Buffalo's situation, giving rise less to understanding of a complex problem and more to a new wave of Buffalo jokes and negative imagery, reassuring readers elsewhere how fortunate they were not to be living there. This concern was hard to build into the research design, but from the start we worried about how we would handle it in the editing and presentation: it became known to us as the "Sunday brunch" problem, in the sense that the magazine article would be a failure if all it did was to present the suffering of Buffalo's people as an object for the detached sympathy of distant, privileged readers—serving it up, so to speak, with the Sunday morning paper and pastry of a sophisticated national audience.

As the work progressed, both of these concerns evaporated. The first proved easy to deal with through the breadth and depth of the selections we made for the article. The second concern proved more or less ungrounded—the people interviewed showed no fear of being exploited, by us or the *Times*, and neither did they dwell on Buffalo and its particular problems. Rather, they had a lot to say about unemployment in both personal and general terms, and seemed glad that someone was finally asking them what they knew about it. In fact, they rather welcomed the opportunity to talk directly to the audience for whom the interviewer was a proxy. Perhaps we could say that they saw themselves sitting at that Sunday brunch table, not as the bagels or coffee cake, a point to which we will return shortly.

The article we submitted to the *Times* had four sections, each introduced by a short paragraph or two of background that we provided. The first section of interviews reflected the breadth of the problem; the second explored Buffalo's particular economic situation and its general setting; the third presented more extensive segments where people discussed the meaning of unemployment in their lives; the fourth offered excerpts in which people reflected on why things were as they were, and what, if anything, could be done about it. This sequence, and the selection and editing of the passages in each section, reflected our assessment of the substance, tone, and drift of the interviews taken as a whole. Although the voices on our tapes were as diverse as the speakers' backgrounds, and although almost every chord was struck at some point or other, our intensive study of the tapes and transcripts showed that people tended to move from personal, localized experience to more general observations; that they spoke in a reflective and analytic mode far more often than in an emotive one; that they addressed their audience directly and even didactically, explaining their experience and what they had concluded from it. All this we tried to capture in our editing and arrangement, a point which should also be held aside, for the moment, with the bagels and cake, for it becomes central to the story at a later stage.

Until this time, our dealings with the *Times* had revealed no differences of purpose or approach. They had fully endorsed our early outline, and immediately accepted the submitted piece for publication. We then entered into the revision and production stage, where variant understandings of the piece began to emerge. Only gradually were we able to distinguish routine editorial differences from more fundamental ones, and so it is worth examining the process in some detail. Our experience can be organized in three major stages, though given the conflict that developed it might be more accurate to think of them as three rounds.

Round One: The first suggested revisions seemed to us quite reasonable; we had been immersed in the research and editing for so long, and so intensely, that we welcomed an outside perspective. The *Times* editors felt that the structure was overly elaborate and academic, and that we needed more explicit, locating discussion about Buffalo to counter the generalized tone of many excerpts. They were not sure they liked the blend of working class and upper middle class voices that we had included, advising more exclusive a focus on working people. In fact, they asked us to submit some additional working class material from our out-takes, more "hard-hitting" testimonies about unemployment for them to choose from.

We were glad to comply. Many in our group, worried that the elite newspaper would undercut the implied politics of a "bottom-up" focus, were reassured by this request for more proletarian testimony, and we quickly submitted the new interviews. . . . We also reorganized the original submission: now it began with a consolidated introductory section putting all of our own comments in one place, set off from the interviews, in order to reduce the academic presence and to allow our subjects to occupy the center of attention in what was, after all, their own article. Their excerpted interviews were now presented in three broad sections—one more Buffalo-specific, one expanding the image of unemployment and exploring its personal dimensions, and a final section, as before, stressing the reflective and analytic voice we had found so distinctive in our research.

We felt the second submission to be a distinct improvement, due in part to the suggestion from the *Times*, and so we were not overly alarmed when, upon receiving it, the editors told us that though we had come in just under the word ceiling they had specified, the piece might still have to be reduced further: this depended on the overall shape of the magazine issue in which we were to appear. We awaited this next round of editing, confident that the piece could remain faithful to our subjects and our research even if it had to be shortened.

Round Two: Shortly after receiving the second version, the *Times* informed us over the phone that they wanted to eliminate entirely the thematic sectional organization of the interviews. They pointed out that it was, after all, artificial in that the interviews tended to overlap. They argued that the various levels and dimensions would come through clearly enough in a montage format. We were unsure of this, but we could not argue strongly against the notion that the words of the people, rather than our intervening structure, should carry the burden of communication. The clinching argument was the *Times'* point that their regular format would be fatal to any other mode of presentation. This is because the article was going to be led by two or three full pages up front, with the rest of the piece marching in single columns through the jungle of high-fashion lingerie ads in the back of the magazine. If we wanted anybody to read it through to the end, they argued compellingly, we would have to capture interest on those opening pages with our most effective material, whatever its place in a thematic arrangement.

They suggested leading with the Rosie Washington interview. Though this is perhaps the most moving and dramatic of our interviews, and one of the longest excerpts for that reason, the choice made us uneasy; we had felt it important in our editing to have the more emotive material set in context, for this is how it had emerged out of far denser and more complex interview transcripts. We worried about leading off with such volatile material, and how this might create a misleading sense of the overall article, obscuring as well the complexity of Ms. Washington's interview itself.

Our uneasiness increased when the *Times'* full rearrangement arrived a day or two later. The order was not the problem—we conceded that the strong lead and the mixed montage were effective, and probably appropriate. But we were concerned about a number of substitutions and omissions that had been made.

In the first place, the *Times* had now moved to eliminate almost entirely the contrast between working class and middle class voices. We could see in the resulting version what we had felt more intuitively in our editing—that the contrast underscored, rather than diluted, the working class focus of the piece, as much through the vivid differences in language as anything else, and that without it a certain social dimension in the interviews was diminished. This is illustrated by the brief but representative excerpts from the Ed Hausner and Gerald Kelly segments, both cut by the *Times*.

But the issues here were hardly major, nor was the impact on the article very substantial. They would have been less bothersome were it not for some other ominous hints in the *Times'* suggested arrangement: they wanted to insert all of the new interviews we had sent them, but to do so they proposed cutting quite a few others, well beyond the few non-working class excerpts we had included. It was at this point that we noticed that virtually all the proposed cuts came from what had been, in our own revised thematic framework, the first and third sections—where people such as Della Love and Stanley Lewandowski discussed and analyzed the Buffalo situation or reflected generally on the meaning of their personal and social situations. But the second section—which tended to emphasize, more descriptively, the personal "impact" of unemployment—was untouched. We also noticed, at this point, that three out of the four new working class interviews accepted by the *Times* were relatively more personal and emotive, and relatively less socially grounded, than the working-class interviews now proposed for omission. That there was an emerging pattern in the *Times'* responses seemed confirmed by a request for an expansion of the Lewis Hawkins interview, much the angriest one we had, a segment close to a *lumpen* cry of pain and a call to violence, quite unrepresentative of the bulk of our collection. Nevertheless, none of these seemed insurmountable differences, and we were reassured by their inclusion of the fourth new interview—"Steelworker"—one that in its control of detail and reflective focus lay close to what we thought to be the heart of the piece. As it happened, however, this was to be the exception that proved the rule, so far as the emerging contrast between our understanding and that of the editors of the *New York Times Magazine*.

Round Three: While we mulled these changes, the *Times* called to say that it had become necessary to cut the article further in order to fit the magazine. But they reassured us that this time it would be possible to retain all the interviews—they had themselves made the limit by eliminating what they called redundant and ineffective verbiage here and there. And with time now a factor, they had also gone ahead and set the piece into proofs. If upon inspecting these we had no objections to the final editing, and they saw no reason to expect any, the article could then move directly to layout and publication.

The excitement of seeing our scrawled-over typescript set into the clean columns of actual print diffused our editorial apprehensions. But not for long—on a first full reading of the new version, the tone of the article seemed strange and unfamiliar. Reading more closely, we discovered why: the new cuts were far from simply occasional prunings. Instead, in one interview after another, crucial sections, often quite extensive, had been excised, in ways that substantially altered the meaning of the words that remained.

In going back and forth between the proofs and the original texts, we began to see a pattern in the *Times'* editorial judgments, a pattern not only responsible for the article's new complexion, but one that threw into sudden perspective the magazine's earlier responses to the material, from the initial call for more "hard hitting" working class interviews to be presented in an almost random montage, to the removal of the middle class viewpoint, to the justified fascination with Rosie Washington and the not quite so justified fascination with Lewis Hawkins. . . . [W]e offer the following reading as a hopefully more than idiosyncratic interpretation of what was going on, informed by our familiarity with the full range of material, and by the insights into the nature of the editorial process obtained in the long struggle to boil mountains of tape and transcript down into a meaningful public representation.

We detected at least three somewhat distinct dimensions to the *Times*'s implicit selection criteria. In the first place, it was clear that in their search for unnecessary verbiage, they tended to settle on passages where people, having made a point or expressed a feeling, went on to explain, support, elaborate, justify, or apply their point. However necessary this might have been editorially, the effect was to undercut or even eliminate the authority with which people spoke, making their statements seem arbitrary and ungrounded, exclamations rather than the products of conscious reflection. This can be seen across class line [*sic*], as in what happened to the Fred Koester interview, or the "Job Counselor" segment. But it is more problematic in working class interviews, of which the Della Love and especially the "Steelworker" excisions—the heart of the interview we had been so reassured to see them accept—are the best examples.

A second dimension to the cuts, is the *Times'* indifference to the self-reflective quality of many of the statements, those passages where the speaker is self-consciously looking at him or herself, often locating that self in a social and class context, implicitly or explicitly. We had been struck by the prevalence of the self-reflective mode in the transcripts, and by its importance for the understanding of what people had to say. The *Times*, however, while approving and in fact encouraging personal statements, seemed to find more reflective self-consciousness to be a kind of unnecessary personal static, interfering with the "real" transmission, and they cut such passages wherever possible. The Bill Phillips and Frank Martinez interviews are good illustrations of this, the latter in particular: this had seemed to us one of the most revealing of all the interviews we had conducted, which is why we included relatively more of it and placed it very carefully near the end, so as to maximize its capacity to draw the article's themes together. The *Times'* editing, however, would have removed almost all of the self-consciousness that makes it a moving expression, one filled with complex social comment. It was in confronting the gutting of the Martinez interview that we finally began to understand that our differences with the *Times* did not trace solely to our journalistic *naivete*, or to matters of taste, but implied rather a fundamentally different sense of what the people had to say, and what it was important for the article to allow them to say.

The third dimension informing the *Times'* editing seemed to us to be their reduction of materials suggesting that the text had in fact been generated in a direct dialogue; this too they tended to see as extraneous to the text-as-statement. By paring the quotes down to the barest bones in the interest of economy, we thought the *Times'* editors had excised or drastically suppressed the definition and expression of personal style, the mode of self-presentation to the interviewer that can embody the complexity of personal identity and social relations in a word or phrase. Such expression, we had found in the transcripts, occurred most often where the interview was most conversational, where the subject addressed the interviewer directly. Because for many good reasons we were not working in a dialogue format, we tried in our editing to capture this sense of style and personal dialogue as the most economical way of suggesting the texture and social sub-text of long, complex interviews.

In this sense, for instance, we had thought the insistent pride and the redundant irony of the Chester Midder selection anything but superfluous; similarly important to us, but not to the *Times*, was the embarrassment that surrounds but does not obstruct Frank Martinez's discussing with a college-educated interviewer his own study of current economics, or Della Love's unembarrassed analysis of Buffalo's decline and potential resurrection, the confident advice of experience offered without any confidence that it is going to be listened to or appreciated. The *Times*, it will be recalled, had earlier sought to eliminate this interview entirely, as well as Stanley Lewandowski's resurrecting of Dr. Townsend's economic prescription, another interview that evokes a sense of communication across substantial social and experiential space. Now the overall editing made it clearer to us that in its search for the essential minimum in each interview, the *Times*, uninterested in the process that had generated the "statements," was in effect flattening to the point of elimination the sense that people were speaking to anyone in particular, much less across a class line clearly sensed and occasionally articulated. Where we saw such references as providing a crucial context, the *Times* saw superfluous asides distracting attention from the basic story each person had to tell.

We hasten to acknowledge that each individual editing decision does not necessarily support the weight of these interpretations; taken together, however, we think the excisions will suggest to readers the patterns we detected. Sculpture, someone once said, is the art of removal, a statue being simply the residue of a myriad of small decisions about what to take away from a block of stone. This well describes the editing of oral transcripts for documentary use, as we found. What the *Times* sought to remove revealed a very different vision of the meaning enclosed in the block of material we had collected: their sculpting would have resulted in the core being emotion and exclamation rather than the reflection and intelligent discussion we found so central in the evidence; it would have emphasized the revelation of experience rather than its instrumental, even didactic communication; and it would have tended to sever that experience from the social and class context with which it had been invested by our subjects, implicitly or explicitly.

This vision seemed to us unacceptable. Fortunately, given how deeply our group felt about this, the *Times* readily accepted our immediate demand to the right of editorship, as long as we managed to meet the word limit they were forced to specify. In fact, they were more puzzled than provoked, claiming to detect, sincerely, we believe, no differences between our ultimate selections and theirs. The final editing then proceeded without incident, though not without hard choices. We restored as much as we could of what seemed absolutely crucial; we cut one or two interviews entirely; and we proved able to find passages that could be cut because the points they had to make were, at least, expressed to some extent or other elsewhere in the article. We were far from satisfied with what this did to a number of interviews, and . . . how much important material still had to be sacrificed. But having earlier become somewhat steeled to this inherent frustration in editing, and accepting the compromises inherent in our particular magazine format, our group felt at the end that the integrity of the editing had been restored, and the larger meaning of the collected material successfully captured in the article.

Lest it seem to readers—as it did to the *Times*—that we were making oral historical mountains out of routine journalistic molehills, a curious coda to our story helps clarify the substantial gulf in sensibility involved, showing as well how this bears directly on the question of the relation between oral history and the exploration of class consciousness.

As finally published, our article bore the title "Down and Out In America." This should seem strange given our argument to this point—nothing we have said about the material or our struggles over its editing suggests that our portrait was of the "down and out." . . . It should not be surprising, therefore, to learn that the title was the original contribution of the *Times*. After tolerating our editorial idiosyncrasies, they announced impatiently that titles were, as an aspect of layout and production, under their sole control. They rejected our best title, "America Not Working," at once a play on the words themselves and a reference to the then current and widely discussed *Working* by Studs Terkel, to whom our efforts owed so much. Admittedly not a perfect title, it had the minor virtue of being consistent with the content of the article. But even after the editorial history just related, we were still a bit staggered by the inappropriateness of the *Times'* creation. It had been crafted, so we were told, by the editor-in-chief, Max Frankel, himself, and was, therefore, unquestionable. The copy editor replied to our protestations by condescendingly pointing out the reference to George Orwell's *Down and Out in Paris and London* as a compensating virtue, apparently indifferent to the fact that this portrait of *lumpen* outcasts was a curious referent indeed for a study supposedly concerned with how working class people perceived the problems they faced as integral members of a troubled community in a troubled society.

The *Times'* fascination with this title crystallized, to us, the values we had sensed in their editing of the documents. For all of the professed concern in commissioning the article, the *Times* seemed comfortable viewing the working class only at a safe moral distance, which their editorial judgments tended consistently to exaggerate. Those who are down and out, to put it simply, do not sit at the breakfast table of those who are up and in. It does not seem unfair to suggest the *Times* sought to offer its readers stimulating fare, not uninvited guests. While they were quite willing to serve up the pain and suffering of the working class, they were less inclined to open their pages to the ideas, values, reflections, advice, and social consciousness of these people. Nor were they interested, to put it another way, in sharing the right to interpret this experience. Tell us what happened and what you feel, the message seemed to be, and our readers will worry about what it means and how to think about it. It is not inconsistent with this spirit to note that the *Times* had been sincerely proud of its intention to include the working class in its magazine. It was the content of that concern that proved problematic, a paradox best summed up in perhaps the single most distressing moment in our relationship with the *Times*, when in the midst of an editorial wrangle the copy editor claimed that the real problem was our inability to provide angrier material. "Can't you give us more stuff," he said, "where, you know, they say how the system's fucking them over?"

The system, indeed. We hope it is not indelicate to suggest that one might learn more about the "system" by studying closely what the *Times* tended to see and not see in these oral historical materials. Surely it is not coincidental that they were drawn to the pain of the working class but not to its subjectivity and consciousness. Nor is it insignificant that they edited their documents so that authority, judgment, and historical self-consciousness tended to remain in the hands of those already controlling the culture, and its mediation through journalism, literature, and history.

* * *

We hope there is more involved here than throwing brickbats at the *Times*, an easy target, after all. We think there are positive lessons in our story, lessons which we began to discover with some surprise, and to understand only gradually. These involve editing as a general matter, but they are especially relevant to those concerned with the study of class, and the special place in this inquiry that oral history is often presumed to have. As we suggested at the start, our struggle, in this sense, was not really with the *Times*, but with the larger problem of understanding how class consciousness can be found in oral history—lower case 'h'—and how as historians and editors we can help it to become visible as we move from the tape to the public table, Sunday brunch or otherwise.

The three dimensions we have described, the self-reflective voice, the social grounding and location of that voice, and the self-conscious engagement of an implied or presumptive audience—these seem to us a good starting point in such a search. As oral historians, we have to be sensitive to these dimensions, however implicit they may be in the tone, stance, and voice of a speaker, or however hidden they may be in the interstices of a conversation. We have to explore the power of this methodology itself, one that is unique in its essential dialogic nature, in its bridging of the historical past and the temporal present, and most fundamentally in its capacity to generate the very documents it then wishes to study.

The lessons can be put in another, more sober way. There is nothing inherent in the oral historical process that guarantees that its documents will be sensitively understood, much less used to create a version of history accurately informed by their unique perspective. The crucial issue is not import, but authority. Those truly interested in a history "from the bottom up," those who feel the limits of the historical reality defined by the powerful, must understand that presuming to "allow" the "inarticulate" to speak is not enough. We must listen, and we must share the responsibility for historical explication and judgment. We must use our skills, our resources, and our privileges to insure that others hear what is being said by those who have always been articulate, but not usually attended to. Only in this way can the arrogance of the powerful be confronted by the truth of another reality, by those history-makers whose consciousness provides the record of that reality and the measure of its challenging power.

Questions for Discussion

1. The following commentary is excerpted from Frisch's and Watts's introduction to their article. Do you think it accurately represents the problems and the interests that were at stake in the case? Why or why not?

> To the *Times*, the issue was simple: the article had to be cut by a few hundred words. But for us, disputes about particular editorial choices crystallized differences that had been implicit almost from the start of the project. This is because we found, and will try to illustrate here, that the material suggested for excision by the *Times* was the material in which people interviewed expressed most directly a larger social and class consciousness.
>
> Such a purge was hardly the conscious intention of the *Times*; surely its editors were sincere in claiming that their only concern was "journalistic" effectiveness. Nor were our differences with them political in the usual sense—in fact, if anything, they insisted that their intention was to produce a more "radical" vision than we felt the interviews supported in their fuller form. But in struggling to restore what we knew to be the heart of the material—although it took us a while to grasp why this was so or what that heart meant—we discovered that the differences were profound, involving explicitly the presentation, and implicitly the existence of working-class consciousness in the material we had collected. We want to suggest that it is far

from coincidental that this issue took the form of judgments about the editing of documents generated by the oral-historical method (Frisch and Watts 1980: 91).

2. Are Frisch and Watts justified in suggesting that the *Times* was *unethical* in its treatment of the Buffalo article? Note that the text quoted above in Question 1 states that the *Times'* way of doing things resulted in an unconscious purge (paragraph 2). If the *Times'* treatment of the manuscript was unethical, how might the problem have been avoided? What does the case suggest about the appropriateness of academicians writing for the popular press?

3. What responsibilities should oral historians have to their subjects beyond assuring that they will be quoted accurately?

4. Is what the residents of Buffalo thought about the article relevant to your analysis of this case?

5. Should the interpretations of the "data" by the interviewing oral historians be given priority over the interpretations of editors? Why or why not?

6. Which of the following editorial changes that the authors allowed the newspaper editors to make do you consider most relevant for assessing whether these oral historians maintained their responsibilities to their subjects:[2]

a. highlighting an emotional interview rather than placing it in the context of "far denser and more complex interview transcripts";

b. eliminating the contrast between working and middle class voices;

c. expanding the material on the "personal 'impact' of unemployment" at the expense of more general, reflective, and interpretive statements;

d. extensively editing the personal statements "in ways that substantially altered the meaning of the words that remained";

e. editing out quotes and comments that personalized the subjects' statements.

Do any of these changes raise moral issues? If so, which ones, and why?

7. In this case, Frisch and his colleagues found themselves caught between contractual obligations to the publisher and their duty to their subjects. What *ethical* issues does this raise? Is it adequately described as a question of choosing either self-interest or protection of subjects? What are a researcher's overriding obligations?

8. How *should* oral history be used and interpreted? Are there fundamental differences between oral history transcripts and other, more traditional sources that affect the way they should be used and interpreted? Explain.

Notes

1. The text has been reprinted, by permission of Greenwood Publishing Group, Inc., Westport, CT, from Michael Frisch and Dorothy L. Watts, "Oral History and the Presenta-

tion of Class Consciousness: The *New York Times* versus the Buffalo Unemployed," *International Journal of Oral History* 1(2): 91–101 (1980). Copyright (c) by The Meckler Corp.

2. The issues in this question are adapted from Frisch and Watts 1980: 94–97.

References

American Historical Association. "Statement on Standards of Professional Conduct." In *Statement on Standards of Professonal Conduct* (Washington, DC: American Historical Association, 1993), p. 1.

American Studies Program. State University at New York at Buffalo. "Down and Out in America." *New York Times Magazine* (February 2, 1975), pp. 9–11, 30, 32–35. (The published article on which the chapter and Frisch and Watts article are based.)

Frisch, Michael, and Watts, Dorothy L. "Oral History and the Presentation of Class Consciousness: The *New York Times* versus the Buffalo Unemployed." *International Journal of Oral History* 1(2): 88–110 (1980).

19 Limiting Access to Scholarly Materials

The Case of the Dead Sea Scrolls

ARE THERE CIRCUMSTANCES in which it is appropriate to restrict access to scholarly materials? If so, what are those circumstances, and what should be the limits of such restrictions? A recent controversy involving ancient documents raises these and other questions.

Part A

Between 1947 and 1956 approximately 800 manuscripts, which were written in the form of scrolls and which dated from about 250 B.C. to A.D. 70, were discovered in the Wadi Qumran territory near the Dead Sea, then controlled by the kingdom of Jordan. About a quarter of these manuscripts—now known as the Dead Sea Scrolls—are books of the Hebrew Bible; the remainder contain sectarian literature that most scholars believe were written by a group of Jews known as Essenes.

These manuscripts are considered to be among the most important documents in the field of biblical studies. They are invaluable for understanding the formative periods of Christianity and Rabbinic Judaism.[1] Only fragments of many of the manuscripts were recovered, some consisting of a single letter.

The initial find, which included seven fairly complete books from the Hebrew Bible, was made by a Bedouin shepherd in 1947 while searching a cave in the Qumran area for buried treasure. Subsequent discoveries were made in 1949, 1951, and 1952 by Bedouins and archaeologists working in the area. The scroll fragments discovered in 1951 were purchased on behalf of the Rockefeller Museum. Further discoveries of manuscript fragments were made during archaeological investigations of the Qumran area in 1953–1956. After the Jordanian government nationalized ownership of the scrolls in 1965, they were housed in the Albright Institute in eastern Jerusalem.

The Jordanian government appointed an editorial board of seven scholars, which had exclusive access to do research on and publish the the contents of scrolls. Each member of the board took responsibility for publishing a certain corpus of scrolls.

When the Israelis annexed eastern Jerusalem in 1967 following the Six-Day War, they took jurisdiction over the Albright Institute and the scrolls. Nearly all of the Dead Sea Scrolls are now housed in Jerusalem's Rockefeller

Museum. A few of the more complete scrolls are on display at the Israel Museum, in western Jerusalem.

The Israeli Antiquities Authority (IAA), which is responsible for the scrolls, allowed the editorial board originally appointed by the Jordanian government to remain in place, although they expanded it to include 13 additional scholars. The IAA also left unchanged Jordan's policy of giving the editorial board exclusive authority over research and publication of the scrolls. The IAA defended the policy of limited access as necessary to protect the rights of the scholars who had already invested considerable amounts of time in assembling and interpreting the scroll fragments. The IAA also justified the policy as necessary for preventing slipshod publications by unqualified researchers.

The Rockefeller Foundation funded a translation project for six years in Israel, after which the members of the editorial board divided up among themselves responsibility for the remaining work on the scrolls and returned to their academic positions.

In the four decades following the appointment of the editorial board, individual members of the board and those to whom the board has given permission have researched and published on part of the corpus of the scrolls. The number of "approved" scholars totals 40. Since their discovery, however, few of the Dead Sea Scroll manuscripts have been published. Rather than publishing reproductions and transcriptions of the manuscripts alone, as a few of the board's (or affiliated) scholars have done, most have been working on interpretive editions, which include commentaries providing the cultural, historical, and literary context for the text. These scholars have spent their entire careers working on these texts. In several cases, members of the team have died and "bequeathed" their rights to colleagues or have "given" the unpublished texts to their students to publish. The next generation of Dead Sea Scroll scholars (i.e., the board members' graduate students) have similarly invested their careers in the publication of interpretive editions of the manuscripts for which they are responsible.

Questions for Discussion

1. Was exclusive access by a board of designated scholars to the Dead Sea Scrolls justifiable? Why or why not?

2. Under what circumstances, if any, would exclusive access to research materials be appropriate? What limits, if any, would you put on restricted access?

3. Assuming some restrictions on access to scholarly research materials are acceptable, what criteria should be used for selecting individuals or groups who will have such access?

4. Applying Question 3 to the facts of this case, what criteria should have

been applied to the selection of the editorial board that was to have exclusive access to these important social and religious texts?

5. The editorial board appointed by Jordan in the 1940s did not include any Jewish scholars. Is this fact relevant to your answer to Questions 1, 3, and 4 above? Why or why not?

6. Who, if anyone, should be considered to "own" discoveries of ancient manuscripts such as the Dead Sea Scrolls? Who should have jurisdiction over such documents?

7. Is there a risk that religious or ideological bias might influence the reconstructions and translations of ancient texts? If so, what is the best way to protect against such bias?

8. Of what relevance, if any, is the American Historical Association's "Statement on Standards of Professional Conduct" (American Historical Association 1993: 1–7) to the ethical dimensions of this case (especially the section headed "Scholarship")? (The text of that section can be found in the Case Notes at the end of this chapter.)

Part B

For many years, biblical scholars waited impatiently for photographs and translations of the reconstructed scrolls to be published. The scrolls were considered to be a valuable source of new insights into the roots of monasticism and Christianity as well as the development of Judaism before the destruction of the Second Temple and the codification of the canon in what is today known as the Hebrew Bible. Finally, in 1988, the editorial board published a limited edition (two dozen copies) of a concordance of the text of the scrolls. (A concordance is an alphabetical list of each word in a document, together with the words immediately surrounding it and noting all the places it occurs in the text. The purpose of a concordance is to assist scholars in locating various portions of a text.) The scrolls' concordance was distributed by the IAA to assist a few, selected scholars engaged in research on the scrolls. Two Bible scholars (who had themselves tried unsuccessfully to gain access to the scrolls) recently used the concordance to develop a computer reconstruction of the scroll manuscripts from which the concordance was made. They also arranged for publication of the reconstructed text, thus making it available to everyone interested in these texts. The chair of the editorial board accused the compilers and publisher of the computer reconstruction of stealing Dead Sea Scroll transcripts by issuing an unauthorized publication.

Questions for Discussion

Leaving aside the question whether the publication constitutes theft as a legal matter:

1. Was the creation of a computer reconstruction of the scrolls from a concordance an immoral theft of the board members' intellectual property? Why or why not?

2. Was the subsequent publication of the computer reconstruction unethical? Why or why not?

3. What factors are relevant to your analysis of Questions 1 and 2 (e.g. the closed access, the length of time the board took to make the manuscripts available, the limited numbers of the concordance that were published, the time and effort invested by the board members in piecing together the scroll fragments which made construction of the concordance possible)?

4. Assuming that a theft occurred, what remedies should be available to the board members and what sanctions should be imposed on the "thieves"?

5. Was it unethical for the two scholars to use the concordance for purposes other than those for which it was published?

Part C

In the intervening years (i.e., between the time the scrolls were discovered and the computer reconstruction was published), the IAA contracted with Elizabeth Hay Bechtel, a philanthropist interested in preservation issues, to have a set of photographic reproductions of the scrolls made. Copies of these photographs would be housed in places other than Israel for safekeeping.

The photography was done, on a free-lance basis, by the Huntington Library's chief photographer. (The Huntington is a private research library located in San Marino, California.) One set of photographs was then deposited with the Ancient Biblical Manuscript Center, an independent research institute in Claremont, California, which Ms. Bechtel had founded and funded. The center was prohibited by contract from allowing anyone to see the negatives without the permission of the scrolls' editors.[2]

After experiencing a falling out with the director of the center, Ms. Bechtel gave the master set of photographs (which she had retained in her possession) to the Huntington Library and donated $90,000 toward the construction of a special vault in which to store them (Wilford 1991: A1). The Huntington had no formal agreement with either Bechtel or the IAA regulating use of and access to the photographs of the reconstructed scrolls.

Ms. Bechtel died in 1987. In September 1991, after the Huntington's new director conducted a review of the library's holdings, the institution decided to grant scholars access to the photographs without restrictions and invited interested scholars to request copies of the photographs for their research. The Huntington's director stated that allowing access was consistent with the library's general policy against imposing restrictions on the use, publication, and reproduction of its ancient manuscripts.

Other librarians and archivists have applauded the decision to allow "open access," although the Dead Sea Scrolls' editorial board opposed the decision, claiming that it could prevent the establishment of a "definitive interpretation" of the texts. The head of the IAA also stated that the original agreement with the institutions holding photographs of the scrolls for safekeeping specified that they were to be in charge only of preserving them and were not permitted to publish them or open access to the public without permission (Haberman 1991: A7). The other institutions holding photographs allowed access only to scholars authorized by the editorial board. Despite its policy of providing open access to the photographs, Huntington officials acknowledge that they are not legally authorized to grant permission to others to publish them.

Questions for Discussion

1. Who should be viewed as having ownership of the photographs of the reconstructed scrolls (e.g., the Israeli government, the members of the editorial board, Ms. Bechtel [or her estate], the Huntington Library, the Huntington's photographer, the Ancient Biblical Manuscript Center, others)?

2. Did the Huntington Library have any ethical obligation to restrict access to the scrolls? If so, what is the basis for the obligation? Of what relevance, if any, are the arrangements between the IAA and Ms. Bechtel to the Huntington's obligations?

3. Should the Huntington have sought the approval of the IAA before it decided to open access to the photographs of the reconstructed scrolls? If so, what is the basis of that obligation?

4. Was it unethical for Ms. Bechtel to give a set of photographs of the scrolls to the Huntington Library without restrictions on their use? Was it unethical for the Huntington to accept the photographs?

5. What role, if any, should academic freedom play in determining the ethics of the Huntington's decision to grant open scholarly access to the photographs?

6. Keeping in mind that the reconstruction of the scrolls by the editorial board required years of piecing together fragments of manuscripts, is unauthorized use of the photographs equivalent to theft of intellectual property? Why or why not?

7. Suppose a scholar wishes to publish research that he or she conducted using the Huntington's photographic copy of the reconstructed manuscripts. What credit does he or she owe to the editorial board for its contribution in creating the reconstruction (without which the research could not be done)? Coauthorship? Acknowledgment? Citation? Nothing?

8. Is acknowledgment or citation enough? What about the board members' claims that the unpublished, pieced-together manuscripts are works in progress? If you think that a coauthorship is more appropriate, what happens if the board member(s) refuse(s) to be an author?

Part D

At its annual meeting in November 1991, the Society of Biblical Literature (SBL)—a 6,000-member association of scholars of the Bible and biblical times—passed a resolution relating to access to ancient manuscript discoveries such as the Dead Sea Scrolls. The SBL's Statement on Access says:

> The Society of Biblical Literature wishes to encourage prompt publication of ancient written materials and ready access to unpublished textual materials. In order to achieve these ends, the society adopts the following guidelines.
>
> 1. *Recommendation to Those Who Own or Control Ancient Written Materials*: Those who own or control ancient written materials should allow all scholars to have access to them. If the condition of the written materials requires that access to them be restricted, arrangements should be made for a facsimile reproduction that will be accessible to all scholars. Although the owners or those in control may choose to authorize one scholar or preferably a team of scholars to prepare an official edition of any given ancient written materials, such authorization should neither preclude access to the written materials by other scholars nor hinder other scholars from publishing their own studies, translations, or editions of the written materials.
>
> 2. *Obligations Entailed By Specially Authorized Editions*: Scholars who are given special authorization to work on official editions of ancient written materials should cooperate with the owners or those in control of the written materials to ensure publication of the edition in an expeditious manner, and they should facilitate access to the written materials by all scholars. If the owners or those in control grant to specially authorized editors any privileges that are unavailable to other scholars, these privileges should by no means include exclusive access to the written materials or facsimile reproductions of them. Furthermore, the owners or those in control should set a reasonable deadline for completion of the envisioned edition (not more than five years after the special authorization is granted).

Questions for Discussion

1. Are these criteria adequate for establishing the rights and responsibilities of scholars assigned to publish ancient texts?

2. If these guidelines had been in effect prior to the discovery of the Dead Sea Scrolls, would they have helped prevent the controversy that later arose?

3. Are guidelines such as these the best way to deal with the problem of access to research materials? Why or why not? What alternatives can you think of that might be more effective?

4. As an ethical matter, should there be a "right" to scholarly access to research materials? What reasons could be given in support of or in opposition to such a right?

Part E

The IAA at first denounced the Huntington's decision as an unethical breach of contract, but by October 27, 1991, only two months later, it agreed to grant "outside" scholars free access to photographs of the scrolls. Although the IAA initially limited access to "personal research only and not for the production of a text edition" (Haberman 1991: A3), by November 26, 1991, it had lifted even this final restriction (Steinfels 1991: A8).

Questions for Discussion

1. Had the IAA retained the restriction on publication of text editions, but allowed other research (including publications other than texts), would you be satisfied?
2. Did the penultimate restrictions comply with the guidelines of the SBL, quoted in Part D?

Case Notes

The following text is excerpted from the American Historical Association's "Statement on Standards of Professional Conduct."[3]

1. Scholarship

Scholarship, the uncovering and exchange of new information and the shaping of interpretations, is basic to the activities of the historical profession. The profession communicates with students in textbooks and classrooms; to other scholars and the general public in books, articles, exhibits, films, and historic sites and structures; and to decision-makers in memoranda and testimony.

Scholars must be not only competent in research and analysis but also cognizant of issues of professional conduct. *Integrity* is one of these issues. It requires an awareness of one's own bias and a readiness to follow sound method and analysis wherever they may lead. It demands disclosure of all significant qualifications of one's arguments. Historians should carefully document their findings and thereafter be prepared to make available to others their sources, evidence, and data, including the documentation they develop through interviews. Historians must not misrepresent evidence or the sources of evidence, must be free of the offense of plagiarism, and must not be indifferent to error or efforts to ignore or conceal it. They should acknowledge the receipt of any financial support, sponsorship, or unique privileges (including privileged access to research material) related to their research, and they should strive to bring the requests and demands of their employers and clients into harmony with the principles of the historical profession. They should also acknowledge assistance received from colleagues, students, and others.

Since historians must have *access to sources*—archival and other—in order to produce reliable history, they have a professional obligation to preserve sources and advocate free, open, equal, and nondiscriminatory access to them, and to avoid actions which might prejudice future access. Historians recognize the appropriateness of some national security and corporate and personal privacy claims but must challenge unnecessary restrictions. They must protect research collections and other historic resources and make those under their control available to other scholars as soon as possible.

Certain kinds of research and conditions attached to employment or to use of records impose obligations to maintain confidentiality, and oral historians often must make promises to interviewees as conditions for interviews. Scholars should honor any pledges made. At the same time, historians should seek definitions of conditions of confidentiality before work begins, press for redefinitions when experience demonstrates the unsatisfactory character of established regulations, and advise their readers of the conditions and rules that govern their work. They also have the obligation to decline to make their services available when policies are unnecessarily restrictive.

As *intellectual diversity* enhances the historical imagination and contributes to the development and vitality of the study of the past, historians should welcome rather than deplore it. When applied with integrity, the political, social, and religious beliefs of historians may inform their historical practice. When historians make interpretations and judgments, they should be careful not to present them in a way that forecloses discussion of alternative interpretations. Historians should be free from institutional and professional penalties for their beliefs and activities, provided they do not misrepresent themselves as speaking for their institutions or their professional organizations.

The bond that grows out of lives committed to the study of history should be evident in the *standards of civility* that govern the conduct of historians in their relations with one another. The preeminent value of all intellectual communities is reasoned discourse—the continuous colloquy among historians of diverse points of view. A commitment to such discourse makes possible the fruitful exchange of views, opinions, and knowledge.

Notes

1. The Qumran documents, which cover the whole Hebrew Bible (except for the Book of Esther), are about 1,000 years older than previously existing versions of the Hebrew text. These documents can assist scholars in tracing the process by which the scriptures developed into their present form (see Vermes 1987: xiv).

2. In addition to the Ancient Biblical Manuscript Center, photographic copies of the

scrolls were housed at Oxford University in England and at Hebrew Union College in Cincinnati.

3. In *Statement on Standards of Professional Conduct* (Washington, DC: American Historical Association, 1993), pp. 1–7. Reprinted with permission.

References

American Historical Association. *Statement on Standards of Professional Conduct.* Washington, DC: American Historical Association, 1993.

Haberman, Clyde. "Israel to Revise Rules on Scrolls." *New York Times* (October 28, 1991), p. A3.

——. "Israel Angry as Library Opens Access to Scrolls." *New York Times* (September 23, 1991), p. A7.

Steinfels, Peter. "Dead Sea Scrolls' Keepers Free Them of the Last Restriction." *New York Times* (November 28, 1991), p. A8.

Vermes, Geza. *The Dead Sea Scrolls in English*, 3d ed. London: Pelican, 1987. (Contains the text of certain nonbiblical texts from the Dead Sea Scrolls, with commentary.)

Wilford, John. "Officials in Israel Ease Stand on Access to Ancient Scrolls." *New York Times* (September 27, 1991), p. A10.

——. "Monopoly over Dead Sea Scrolls Is Ended." *New York Times* (September 22, 1991), pp. A1, 14.

20 Faculty–Graduate Student Relations

The interaction between and among academic colleagues is a core element of intellectual interchange and growth. The academic environment therefore provides a forum in which scholars and budding scholars can exchange ideas. As we have seen in many of the cases, academic relationships are anything but free of ethical difficulties. The relationship between faculty and graduate students is, however, even more complex than the relationship between already-established scholars. The dependency of the student upon the faculty member is obvious; the power imbalance is equally obvious. Add to the dependency and power imbalance the ambiguity of the student-as-author who is writing for courses, sharing his or her ideas with faculty, working for faculty as a research assistant, and, perhaps, writing his or her own works for publication, and you have the potential for significant problems.

By virtue of his or her status as a student, the graduate student may not be considered a "colleague," to whom certain rights naturally accrue. On the other hand, many of the students' interactions with faculty are "collegial" or "colleague-like." The problem is both to understand the complexity of the relationship and to come to terms with the rights and responsibilities of both students and faculty toward one another.

This set of cases is designed to explore some of the problems associated with the graduate student-as-author. To what extent are the faculty in these cases pursuing legitimate pedagogical goals using well-accepted (and ethically unproblematic) techniques? What intellectual property rights do graduate students have with respect to their work? To what extent is the work of the various students in these cases the work of a "colleague"? What justifications can appropriately be offered to distinguish the work of the graduate students in these cases from the work of professionals (i.e., had the faculty member in each case had to collaborate with a colleague in order to do the same work)? What is the proper relationship between mentor and student?

■ Case 1

Linda is a third-year graduate student in U.S. military history. She is disturbed by something that happened in yesterday's seminar on the Vietnam era and decides to talk about it with her classmate George. "In most of my classes, the professor has us come and see him about a paper topic. But yesterday, in Professor Crane's class, he handed out descriptions of what he wanted each of us to write about. They are clearly chapters of a book he is planning to write. This is so weird, I can't believe it. He's a great professor and everything; lots of people come here to work with him. But first of all, I don't think we should be doing his research for him, and second of all, the topic he gave me is something I'm not at all interested in. Have you ever had this happen to you?"

"No," answers George, "I haven't. Why don't you go and talk to him?"

Questions for Discussion

1. If you were Linda's friend, what additional information, if any, would you want to know before advising her? What would you want her to ask Professor Crane?

2. Assuming that Professor Crane does intend to have the students in the seminar research and write on the subjects he plans to include as chapters of his book, what ethical concerns, if any, does such an arrangement raise?

3. If requiring students to research and write on particular subjects for the professor's benefit raises ethical questions, what actions should (and can) Linda take?

4. What role, if any, might academic codes play in resolving Linda's problem?

5. What difference, if any, would it make if all the students received a "research stipend" of $1,000 for this research and writing?

■ Case 2

Stan is a graduate student in Russian history. His command of Russian and his research and analytical skills are excellent, attested to by the high level of his academic work and the fact that several professors are interested in hiring him as a research assistant. Stan decides to accept Professor Grebe's offer to do some research work for him. Grebe works on Imperial Russia; his current project is on the secret police under Alexander III. Stan's assignments include reading through a run of periodicals looking for any and all references to the secret police, plus any materials he (Stan) considers related to the topic. He will take verbatim notes of whatever relevant materials he finds. Stan will be paid $9,000 for the year's work (the standard research assistant salary at his university), and Professor Grebe will acknowledge Stan's contribution at the beginning of any publications that result from this project, where his work is related to that part of the project.

Questions for Discussion

1. What ethical concerns, if any, does Professor Grebe's arrangement with Stan raise?

2. Identify the elements of research work that Stan is providing Professor Grebe. To what degree does Stan's work constitute interpretation of historical materials? What will Stan provide to Professor Grebe, and how will Professor Grebe use what Stan gives him?

3. How does this case differ from Case 1? Of what relevance is the fact that Stan will be paid for his work and that Linda is enrolled in Professor Crane's class and doing the work for course credit? Of what relevance is Professor Grebe's acknowledgment of Stan's contribution? Do these distinctions (and others that you have identified) make a difference to your analysis, or are they distinctions without a difference?

■ Case 3

Linda, the graduate student in military history from Case 1, has now completed her course work, passed her qualifying exams, and is putting together the prospectus for her dissertation. The chair of her committee, and the person with whom she works most closely, is Professor Henriette Tern. Linda came to the university in part because she wanted to work with Professor Tern, and has been happy that the professor has taken on the role of mentor. Their relationship has been a good one. Linda's own methodological ideas and research interests in many ways resemble those of her mentor, perhaps as a result of having worked with her. But now, Linda is troubled. Professor Tern seems to be directing Linda toward topics that fill in gaps in her own work, while Linda wants to pursue some different approaches. Professor Tern studies the military history of the South in the Civil War; her first two books are broad surveys of the role of the rebel troops from various states. She wants Linda to work on tactical aspects of the use of the rebel troops from Alabama, while Linda wants to study questions of leadership, culture, and the social origins of various military units. In some ways, in fact, Linda's ideas conflict with Professor Tern's research. She feels that she has some important contributions to make to the field, and is afraid of becoming either a clone or an epigone. Having come this far, however, Linda is nervous about rocking the boat.

Questions for Discussion

1. Does Linda have anything to feel nervous about? Are her fears justified? Has she misread Professor Tern's intent? What steps can she take to deal with the situation?

2. What is the proper relationship between mentor and student? Has Professor Tern overstepped the limits of that relationship?

■ Case 4

Bob was in Professor Marsh's seminar on historiography last year and presented a paper that was the source of much heated discussion. In the end, however, it was clear that the discussion resulted in a very useful exchange of ideas, for which Professor Marsh expressed his appreciation. Bob's research took him to some unusual sources, which contributed to his unique approach to the subject.

Professor Marsh has been invited to give a paper on the influence of Darwin on U.S. historical writing, and decides to take up some of Bob's controversial historiographical ideas. He starts with the unusual sources cited in Bob's footnotes. The paper is well received, and Professor Marsh decides to publish it, acknowledging Bob's contribution in a broadly worded footnote at the beginning of the article.

Two years later, Bob sees Marsh's article in the *Journal of High Ideas.* At first he is delighted to see that Marsh has thought enough of his ideas to use them and to acknowledge him in a note. Then he wonders, "If the ideas were so good, why didn't he suggest that I work on them for publication?"

Questions for Discussion

1. Has Professor Marsh done anything inappropriate? If so, what? Of what relevance is Marsh's acknowledgment of Bob's contribution?

2. Would it make a difference if instead of "mining" the footnotes of a graduate student's paper, a professor "mined" the footnotes of a published book or article?

3. What obligations do faculty have to promote their student's work? How long should a professor wait to see whether a student is going to publish before going forward with ideas the student generated?

4. How would it change the case if Marsh were dealing with a colleague rather than a student?

5. How might Marsh have used Bob's work in a way that would be fair/satisfactory to both of them?

■ Case 5

Sam, a graduate student in history, has worked with Professor Brooks as a teaching assistant in three courses and is now a research assistant for her. They have a very good working relationship even though their central interests are quite different; in fact, Professor Brooks is not even on Sam's dissertation committee.

Professor Brooks receives a grant to edit a collection of essays on the history of the American South. She envisions dividing the collection into three main sections and writing a general introduction and an introduction to each of the sections. She knows that Sam really is more of an expert on the topic of one of the sections than she is. Sam's workload as Brooks's research assistant has been light for several months, whereas Brooks's workload has been quite heavy, so she asks Sam to write a draft of an introduction to the third section. "Just jot something out quickly," she says, "to help me organize my thoughts."

Sam knows that it will be easy for him to "jot out" a draft introduction, and he does so over the next two weeks.

Questions for Discussion

1. Is it appropriate for Professor Brooks to give Sam this assignment? Is it less appropriate to ask a research assistant to *write* something than it is to ask her or him to *research* something?

2. Suppose Professor Brooks likes Sam's draft very much; she rearranges it a bit, adds a section, drops a few paragraphs, and generally polishes the writing. Then she uses it as the introduction, citing Sam only as her research assistant in the general acknowledgements, without making reference to his written contribution. Is this okay? How much of an acknowledgement should Sam get?

3. Suppose Sam's draft is so good that Professor Brooks hardly changes a word. How much of an acknowledgement should he get?

4. Now imagine that rather than writing an introduction to one-third of the book, Sam writes a draft of the general introduction. Due to time constraints, Professor Brooks decides not to include the introductions to the separate sections. Now it turns out that Professor Brooks's contribution to the book amounts to making the difficult decisions on what essays to include, getting the grant, and asking Sam to write the introduction. How much of an acknowledgement does Sam deserve in this case?

21 Intellectual Property

THE SUCCESS OF the scholarly enterprise depends both on the exchange of ideas and on the ability of the academic community to judge the quality of the scholarship being produced. These two aspects of the academy are often intertwined: papers are presented at conferences; faculty review each other's work as part of the peer review process; the corpus of scholarship builds upon itself—in this way each scholar is dependent on and indebted to the work of every scholar who has gone before him or her.

The cases in this chapter explore the ethical issues raised by some scholarly practices that may be seen as impediments to the openness critical to an effective academic system. In particular, the status of the unpublished dissertation is examined, as is the practice of "footnote mining" (Cases 1 through 3 and 5). One case (Case 6) deals with intellectual property rights in newly discovered documents, and another (Case 7) with the peer review process.

■ Case 1

Professor A is the editor of a major historical journal. One of her associate editors recently brought to her attention a submission that raised questions about the editorial policy of the journal. The associate editor, Professor B, was concerned because the review article submitted by Professor C included a lengthy essay about an unpublished dissertation.

"It's true that we don't ask reviewers to include doctoral dissertations in review essays," said A. "But I'd think that the author of the dissertation would be tickled pink that C has written about his work in such detail," she said. "Professor C is an excellent scholar. The fact that she included his dissertation with the other two books should please him."

"Well, I don't think it's right. If everyone knows about his dissertation before he's had a chance to publish it, someone's going to gut it. What's to stop some senior person working in the field from 'mining' his notes?"

"But the dissertation is available through University Microfilms, there for anyone who wants to read it."

"It's not the same thing," said B, shaking her head. "Dissertations have to be really looked for, but writing about them in review essays notifies everyone, including unscrupulous scholars—there are some, you know! If someone else publishes first, using his sources, no top-rate press will want to publish the dissertation. He'll be preempted."

Questions for Discussion

1. Should journals review doctoral dissertations? What ethical concerns does inclusion of dissertations raise?

2. Are B's concerns about preemption justified, or is this a nonissue?

3. If B's concerns are valid, what should the journal do (assuming the essay is otherwise acceptable for publication in the journal)?

a. Reject the review essay outright.

b. Ask the author of the review essay to rewrite it, omitting discussion of the dissertation.

c. Seek permission from the author of the dissertation before accepting the review essay for publication.

d. Other (explain).

■ Case 2

Professor Jones just published his third book, *Fields of Dreams*, on land reform in twentieth-century South America. Chapter four relies heavily on a doctoral dissertation whose author, Dunn, is now a second-year assistant professor at a major research university. Dunn is in the process of revising the dissertation into a book; he intends to submit the manuscript to the top publisher in his field next year.

Jones properly attributed the ideas presented in chapter four to Dunn in an initial footnote. When Dunn read Jones's book, however, his reaction was one of shock. "Who will want to publish my book now? All of my best ideas are in Jones's book. No one is going to read the notes, and even if they do, the fact is that Jones said it first. It's books that count, not dissertations."

Questions for Discussion

1. Has Jones done anything wrong? What use can scholars ethically make of unpublished dissertations?

2. Are Dunn's fears justified?

3. Would it make a difference if instead of "relying heavily" on Dunn's dissertation, Jones simply cited it for several specific points? Why or why not?

4. Would it make a difference if Dunn's work were a book or an article rather than a dissertation? Why or why not?

5. Would it make a difference if Jones heavily relied not just on chapter four of Dunn's thesis, but also on chapters two, three, and five (but not the introduction or chapter six)? Why or why not?

6. Suppose that eight years have passed since Dunn completed his disser-

tation; he has not yet published a book or major article based on it. Suppose also that Jones cites the dissertation appropriately throughout his work. Can other scholars treat the dissertation as if it were a book? How long should they wait to see if he will publish it? Must his permission be sought before using it? What if the dissertation were 20 years old?

■ Case 3

Dunn (from Case 2) decides to talk to his colleague Blane, a fifth-year assistant professor whose book came out last year. "Did anything like this happen to you?" asked Dunn. "I mean, he didn't just cite me, recognizing that my work was out there, or cite some specific points I make. He gutted my thesis!"

"No, nothing like that could have happened to me," responded Blane. "You were at Yale, right?" she asked. "Well, at Harvard, they have a policy of withholding dissertations for five years. I could have released it, but I figured that the protection against getting scooped was more important than being 'of record.' I knew I'd eventually send it to University Microfilms, but by then my book would be out, or close to it."

Dunn, looking puzzled, asked, "Didn't anyone working on a similar topic get in touch with you and ask to see it? Also, you must have gotten feedback on the chapters from people outside your thesis committee."

"I've just been very guarded," explained Blane. "A couple of times I was asked outright, and then I relied on Harvard's policy—I told them that Harvard has a five-year hold on it."

Questions for Discussion

1. Although the policy of withholding dissertations from distribution used to be more widespread than it is today, it has not disappeared; several universities continue to have such a policy, under which scholars who want access to dissertations must get permission from both the author and the university. The justifications are that the author has a privilege to hold on to his or her own work and that the university has a right to protect the author, assisting him or her to exercise this privilege.

Are such policies ethical? If so, are there limits to the length of time that dissertations can ethically be withheld? Do such policies give their students an unfair advantage?

2. Are dissertations the same in terms of intellectual property as published books or articles? How are they the same and how are they different? Do the differences raise special concerns or call for special treatment?

3. How can junior faculty protect their dissertations from being "mined," "gutted," or otherwise misused and still be collegial?

Case 4

Jane Doe, a British historian, wrote her dissertation on a major nineteenth-century journalist and radical. Her work was based on a significant amount of archival research. However, Doe has refused to make her dissertation available to other scholars. It has now been many years since the dissertation was written; Doe has published numerous works, none of which deal with the subject matter covered in the dissertation. It is clear that she will not, in the remainder of her career, publish on the central figure in the thesis, but that other scholars continue to be interested in him and would like to see her work.

Questions for Discussion

1. Does Doe have any obligation to make her dissertation available to other scholars?

2. What is the nature of a dissertation? Is it simply a manuscript, of which the author has the sole right of control (to burn, if he or she wants to), or does the fact that its production is a requirement for obtaining a doctoral degree affect the nature of the document and therefore also affect the right of control?

3. Professor Smith is working in the same area as Doe's dissertation topic, but can't get her to let him see the thesis. Is his only choice to do the same archival work himself? What if Doe has, through the years, stated that she intends to publish it some day? What if Doe finally decides to go ahead and publish the dissertation while Smith is in the midst of his project?

4. What if the materials Doe worked with were private letters and documents, to which she was given exclusive access, or were otherwise difficult to gain access to?

Case 5

Dunn (from Case 2) has another friend, Edwards, who has had a similar, though different, experience. Edwards's dissertation was on nineteenth-century Georgia. He has been revising his dissertation and hopes to publish it with a major university press. Thomas, a scholar who, though in the same general field as Edwards, works on a different topic, has just published a book. Thomas's new book, on farming in the South before the Civil War, departs from his usual line of research. Edwards reads the book and finds to his initial satisfaction that his work has been cited. On reading further, however, he finds that his dissertation was cited only once, for a rather obscure point, but that Thomas has extensively cited the sources Edwards used. Edwards's feeling is that Thomas simply "mined" his thesis for sources, but only gave him credit for the one detail. He thinks that, at best, Thomas went back to the archives to do the research independently, using Edwards's sources as a starting point;

and at worst, Thomas stole Edwards's work, citing sources he had never seen. In either case, Edwards thinks, Thomas should have cited him as his source.

Questions for Discussion

1. Did Thomas do anything wrong? If so, what?

2. Is it relevant that Edwards's work is a dissertation rather than a published book or article? Why or why not?

3. What, if anything, can Edwards do?

■ Case 6

Clark is working on a project on the relationship between army and state in Weimar Germany. While working in the archive, he comes across several boxes of uncataloged documents. Browsing through them, he is amazed to find that they are documents from several important military leaders, including private diaries, the existence of which is, to the best of his knowledge, unknown. His future work with the newly found documents reveals that they are a treasure trove of valuable information.

Questions for Discussion

1. What obligation does Clark have to let other historians know about the existence of these documents?

2. Does or should Clark have some kind of right of limited access to them? If so, how long should other scholars have to wait to gain access to them?

■ Case 7

Professor Q has been asked to referee three articles for publication in three different scholarly journals. Reviews for these journals are not masked: she knows who wrote the articles, and the authors will know who reviewed them.[1] Professor Q is a fairly junior scholar at a well-respected research university. The author of the first article, Professor R, is of about the same rank as Q, at a similar institution.

The author of the second article, Professor S, is a senior scholar who holds an endowed chair at Yale and is known to hold a grudge.

The author of the third article, Professor T, is a senior scholar who has not published in some time, but is beginning to make a comeback.

Questions for Discussion

1. Despite the fact that all reviews should be written fairly and honestly, what difference might the identities and characteristics of the referee and authors make to Professor Q's review? What if Professor Q were a well-re-

spected senior scholar, author of several seminal works in her field (and that Professor R is a junior scholar)?

2. Would masking affect Q's ability to write fair and honest reviews? In what way(s)? (Consider both single- and double-masking.)

3. What benefits and drawbacks does masking have from the referee's perspective? What benefits and drawbacks does masking have from the author's perspective?

Note

1. The review process is often "masked"; that is, either the author, the referee, or both do not know each other's identity. If neither knows the other's identity, the review is "double-masked"; if one or the other knows, the review is "single-masked." Another term for "masked review" is "blind review."

Instructional Notes

4. Scientific Misconduct: What Is It and How Is It Investigated?

The events of this case are presented in a multipart format, with suggested questions for discussion following each portion of the tale. You may wish to use all of these, or to select only a subset for each portion. You can also readily eliminate one or more sets of questions by linking segments of the narrative without the intervening questions. Or you can choose to use only parts of the narrative as shorter cases. The point is that you can easily tailor the discussion of this case for your particular group and the issues you wish to address.

Discussion of this case will raise issues concerning the accepted standards of conduct in your laboratory and in your area of research. Among the topics that will come up are how laboratory notebooks should be kept and who participates in the preparation of a manuscript for publication. You might also wish to discuss such issues as the most beneficial function of:

1. lab group meetings;
2. presentation of results ready for publication;
3. dissemination of results in the form of compiled and derived data; and
4. group discussion of primary data and the conclusions that can be drawn from them.

Discussion of this case could be a vehicle for the presentation and discussion of the research ethics policies of your institution and of its procedures for investigating alleged scientific misconduct. In addition to the two references already cited, you may wish to consult the following sources:

National Academy of Sciences. *On Being a Scientist.* Washington, DC: National Academy Press, 1989.

U.S. Department of Health and Human Services. Public Health Service. Office of the Assistant Secretary for Health; Office of Health Planning and Evaluation; and Office of Scientific Integrity Review. *Data Management in Biomedical Research: Report of a Workshop* (April 1990).

5. Authorship and the Use of Scientific Data

The cases in this chapter have been designed for use in an introductory undergraduate biology course in which the students are likely to be freshmen and/or nonmajors. They are probably most appropriate for a class that includes either a laboratory or discussion section led by graduate student assistant instructors. For this reason, the cases are fairly straightforward and not highly problematic. However, they do include cases that involve ethical problems confronting professionals as well as students in an effort to illustrate to the students that they are a part of the scientific enterprise.

The authors of these materials assume that students will have read the cases before class and will be prepared to discuss their answers to the questions for discussion and similar questions posed during class. Only a few questions are posed at the end of each case. The intention is to start students thinking about the case ahead of time, but not to have them consider all the issues raised by a case before the discussion. Additional questions are suggested below in these Instructional Notes, and you may wish to have the students consider some of these questions before class as well.

As stated in the introduction to the chapter, these cases use a casuistic approach to moral problem solving, which involves reasoning by analogy from earlier cases to reach solutions to later ones. However, the cases do not require this approach to be useful, and thus do not need to be used as a complete set. Instead, they may be used individually or in small groups, depending on the instructor's time and inclination.

The following notes are designed to provide some suggestions to the instructor on points that could be raised during the discussion of the cases.

Cases 1 and 2

These should be easy cases on which to reach consensus. They are intended to be paradigms on which further discussions can be based. Do not spend too much time on these. Rather, try to establish a consensus that the actions of Atos in Case 1 are clearly wrong, and that those of Bishara in case 2 are right. But do point out that in Case 1, Atos's plagiarism of another's thesis is wrong both for scientific reasons (because the data might be bad), as well as ethical ones (passing off another's work as one's own is dishonest, and diminishes the principles of trust and integrity on which the scientific enterprise is built). Further, there is no relationship between the two students that might suggest that Atos had permission to use the other's work.

In Case 2, you should mention that presentation of limited amounts of others' work in a publication is proper if required for coherence of the presentation, and if permission is obtained and proper citation made. In part, what is acceptable in using another's work depends upon custom and practice from discipline to discipline, and even from lab to lab. If time permits, you may want to change the scenario to one in which the lab director is trying to coerce Bishara into allowing him to use her work in his own paper. If the director has permission to use Bishara's work (albeit coerced) and cites it properly, his use would be "correct," but most scientists would still consider it unethical.

Case 3

In the discussion of Case 3, students should see that, although there are some similarities to Case 2, this case is more like Case 1. Copying something and presenting it as yours is always wrong. Chan may not have stolen the report out of his roommate's drawer, but neither was he given permission to copy it. In any event, one cannot obtain permission to do wrong (here, presenting his roommate's work as his own). Therefore, we can conclude that both citation and permission are lacking in Case 3, and that it is most similar to Case 1. Chan's action also represents a breach of implied contract by the researcher with the profession to do original work. Students doing class experiments are also part of the scientific endeavor although they may not see themselves in this way. Frequently, students have the idea that high ethical standards somehow switch on when one becomes a professional researcher or physician. To help them see that what they do now is of consequence, it may help to ask them whether they would seek out the services of a physician who they knew cheated in an undergraduate lab course they had taken together.

In order to stimulate discussion, you may wish to utilize some of the following questions:

1. Was something stolen? If so, what?
2. What would have been the proper course of action for Chan to have taken in this situation?
3. Would it have been all right if Chan had only used his roommate's data and had written his own report?
4. Would your evaluation of this case be different if Chan had missed the laboratory to attend his grandmother's funeral? Or because he had been involved in an automobile accident?

This case gives you, as the instructor, an opportunity to articulate what students in your course should do if they find themselves in a situation like one of these scenarios. You can appropriately use this opportunity to discuss your policies regarding missed labs as well as consequences for plagiarism.

Case 4

This case is a bit trickier. Dunn has not directly done any experimental work, so whether the data are legitimately his to use for his report depends on whether the protocol was part of the assignment. Instructors will have different policies, depending on their pedagogical objectives and whether part of the purpose of assigning laboratory work is to develop students' manual and technical skills. Although not the best way to learn scientific techniques, the actions of Dunn and his lab partner would be considered by most scientists to be ethically acceptable. Dunn was present and participated in the experiment even if his participation was more visual than manual. This assessment assumes that the pre-lab protocol was not a graded assignment that Dunn's partner copied, and that evaluation of manual dexterity was not an objective of the course. Here is an opportunity to discuss the objectives of your course and the standards by which the students will be evaluated.

In this case, since Dunn has a partnership with his lab associate, he has implicit permission both to use and cite the data. However, it would not have been proper for Dunn to prepare the interpretive portions of the reports and for his partner simply to modify them before submission.

Possible questions for discussion:

1. How does this case differ from a modification of Case 3 in which Chan only copies his roommate's data? Do you consider these differences significant?

2. Would this arrangement be ethical if the pre-lab protocol were a graded assignment that Dunn's partner copies before handing in?

3. Would it be ethical for the arrangement to include an agreement that Dunn write the first draft of the lab report, which his partner then uses to write her report?

Case 5

This case presents another version of the scenario involving the questionable use of another's data. In this case, Estoban has done the work expected of her, and more. Yet for unknown reasons, which may not be her fault, she has a data set that prevents her from preparing the assigned laboratory report. The question becomes whether her use of another person's data is justified. Her actions suggest that both she and Dunn feel that it is not, as they decide that a modification of the data is required.

This case has similarities to Cases 2 and 4, but also some important differences. Students should be encouraged to explain why they have chosen a preceding case as the most similar and why they consider any differences as significant or insignificant. In addition, students should recognize the importance of respecting the integrity of the data.

We would suggest that Case 2 is the closest to this case, because unlike Dunn's partner in Case 4, Estoban did not participate in generating Dunn's data (although she does have permission to use them). *However*, it is significant that Estoban altered the data and did not cite Dunn as the source of the data. These differences force us to conclude that Estoban's actions are unethical. If Estoban had reported her own data, possibly adding her hypothesis as to why the data were not as expected, and then cited and used Dunn's data set as-is, her actions could be considered ethically sound.

How do you, as an instructor, treat similar situations? Do you penalize students for "poor" data? Would you have penalized Estoban for citing and using another student's data as suggested in the preceding paragraph? Some would argue that only outstanding students can write a good report based on "poor" data, while most students can produce a good report with "good" data. To try to address this concern, many instructors now have their students pool their data and write their lab reports using the pooled class data.

Suggested discussion questions:

1. What criteria did you use to determine whether Estoban's actions were ethical?

2. What should Estoban have done in this situation?

3. What are the differences between this case and Case 2? Do you consider these differences significant; that is, do they force you to conclude that Estoban's actions were unethical while Bishara's in Case 2 were ethical?

4. What are the differences between this case and Case 4? Are they significant?

5. Do ethical considerations really matter in an undergraduate laboratory such as this one, where the same experiments are done each semester, and thus no new information is added to the body of scientific knowledge? Why or why not?

Case 6

With this case, we move into the realm of professional science. Most students will be more willing to condemn Professor Field than the students of Cases 3, 4, and 5, yet the issues are the same. For this reason, you may wish to discuss this case between Cases 4 and 5 rather than after them. As noted before, cases involving students have been juxtaposed with cases involving scientists so that students will discover that they are part of the scientific enterprise and that research ethics don't suddenly switch on when one becomes a professional scientist; they must be part of the way one does science from the beginning.

Some possible discussion questions:

1. Do you find another case more similar to this one than Cases 1 or 4?

2. Are there ethically significant differences between this case and Case 4? (Field did come up with the idea for the experiments independently. That is, she did put intellectual effort into the experiments and into the analysis of the data; *but* she did not participate in them and has no permission to use the data, nor does she cite the previous work. These are important differences.)

3. What would have been the proper course of action for Professor Field to have taken?

4. Would your answer differ if Professor Field's tenure depended upon this publication? (The "publish or perish" mandate is a common explanation given

for scientific misconduct [see Campbell 1987; U.S. House of Representatives Subcommittee on Investigations and Oversight 1990: 3].)

5. Would you alter your evaluation if the Hungarian author were alive?

6. If Professor Field has already carried out her experiment when she discovers the previous research, would it be acceptable for her to publish her work without reference to the Hungarian publication? (Note that one cannot publish research that duplicates previously published results; Field would receive no credit for her idea and labor.)

7. Does the subject of the research affect the seriousness of the ethical breach, making it either more or less serious? For instance, do Professor Field's actions matter, since the influence of music on plant growth is not a critical area of scientific inquiry? Would it be different if the research had to do with a cure for AIDS?

8. If Professor Field's actions are discovered by a colleague, what consequences should result? Some possibilities are:

a. no penalty could be imposed, at least for a first offense;
b. if untenured, Field could be denied tenure;
c. the Hungarian publisher could sue Field;
d. Field could be required to take a leave from the university without pay;
e. the university could fire Field;
f. Field could be ostracized by the community of scholars of which she has been a part;
g. the article could be retracted from the journal in which it was published;
h. Field could be denied federal funding for some number of years.

It may be useful to discuss standards of evaluation. Should the conduct of professional scientists be evaluated in the same way as that of students? Should the penalties be similar?

Case 7

This case adds some extra layers of complexity to questions of authorship in the form of the conventions of data ownership in a biology research laboratory. You, the instructor, will have a greater understanding, and possibly acceptance, of the way things usually are done. It may be best to let the students discuss and justify their evaluation of this case without too much input from you at first. Try to avoid taking the approach that this is how scientific research works in the real world and so there is little to discuss other than the lack of citation. Most professional scientists would argue that if Johanson had been cited in either the text or in the acknowledgements, Ghana's and Hiromoto's actions would have been ethical, even though Johanson was not consulted or given a coauthorship.

Some questions to use in discussion:

1. What cases did you use to reach your decision regarding the events in this case?

2. Were there any differences between this case and previous cases that you had to take into consideration?

3. If you were Ghana and concluded that some of Professor Hiromoto's actions were improper, what would you do? Is the impropriety significant enough to warrant taking some action? How would you go about questioning your research advisor or employer?

4. If Professor Hiromoto's actions are discovered by a colleague, what consequences should result?

5. Would you alter your evaluation of this case:

 a. if Johanson's work had been done in Hiromoto's laboratory and under her supervision but was funded by a grant awarded to Johanson?
 b. if the research was funded by Hiromoto's grant and carried out in her laboratory, but Hiromoto had essentially no intellectual input because of the friction between her and Johanson?
 c. if Ghana could not repeat Johanson's experiments for at least a year, if ever, because of a freezer failure resulting in the loss of essential research materials?
 d. if the research were in the field of DNA replication, and the model supported was new but not earth-shaking?
 e. if the research were in the field of cancer treatment and demonstrated that a new mode of treatment could save lives?

If you would change your opinion given any of these alternatives, what is it that you feel justifies the change?

6. Data Alteration in Scientific Research

These cases represent a continuation of the set of cases in chapter 5 targeted for use in introductory undergraduate biology classes.

As before, the authors suggest that the students using these cases read them ahead of time and prepare answers to the questions posed at the conclusion of each case. They should be instructed to be prepared to participate in a discussion of these and similar cases during class or discussion period.

As stated in the introduction to the chapter, these cases use a casuistical approach to moral problem solving, which involves reasoning by analogy from earlier cases to reach solutions to later ones. However, the cases do not require this approach in order to be useful, and thus do not need to be used as a complete set. Instead, they may be used individually or in different groupings, depending on the instructor's time and inclination. One would, however, need to modify the questions if a casuistic approach were not used. For instance, by posing different questions, these cases can be used to focus the discussion on the official definition of scientific misconduct and various questionable research practices. You may also wish to use the discussion of these cases as an opportunity to inform students of your institution's procedures for reporting and investigating allegations of misconduct.

What follows are some suggestions for teaching the cases as well as follow-up questions for each case.

The first four cases present variations on the theme of data alteration. The next three cases, 5 through 7, deal with the omission of data. Case 8 is a more difficult case in which the difference between data alteration and interpretation must be explored. It is based very loosely on some of the issues raised in a recent misconduct case.

Cases 1 and 2 and General Issues

It is probably best to discuss these two cases together. Case 1 is intended to describe behavior that is clearly ethically bankrupt, but Able and Baker are cast as otherwise model students. In Case 2, Chase is a less than admirable student (the type that drives lab instructors crazy), but his actions in reporting his data are sound. These characterizations are intentional because we tend to criticize and scrutinize the students who fit the "mold" much less than we do those who do not fit as well. Ethics has nothing to do with fitting the "mold."

The primary objective of these cases is to facilitate discussion of what are proper and appropriate research ethics. Another important objective, however, is to impress upon students that they are part of the scientific enterprise even if it is "only a class." Research ethics are important throughout science. In the course of discussing these cases, it may help to have the class discuss what would happen if odd data were routinely altered or omitted by professional scientists. (Examples using professional scientists are presented in the later cases.) Proposing that students imagine such behavior occurring in the field of cancer research or some other health-related field may make the consequences more personal and relevant to the students' lives. You may wish to note, for instance, such consequences as the missed opportunities to improve our theories, the failure to discover something new and unexpected, or the inability to reevaluate the data at a later time.

Another objective is to raise the awareness of students and instructors of how they can change "the system" to be more concerned with ethical issues. Using these cases will raise questions concerning what you, the instructor, expect of the students. What are your ethical standards for the course? How will you evaluate students and their reports? Is correctness of the data important? Is advance preparation for labs part of the course requirements that will be graded? You should be very clear about your standards and criteria and be prepared to address students' questions concerning their appropriateness. You may decide to address these two secondary objectives now, with the first two cases, or to wait until later in this series of cases when the students may bring them up on their own.

The following are additional questions to use in your discussions of the first two cases:

1. Do you have any objections to this approach to experimental science in general, for instance, in cancer research? If your answer to this question is different for Case 2 than for Case 1, what criteria do you use to differentiate the two?

2. How do you feel about the behavior of the professor teaching this course?

3. What would you like to see done to improve the course for everyone enrolled? How could students effect such changes?

Case 3

Donna Donovan's alteration of data to make the graph prettier and the report easier to write is something many students have done at one time or another. Most students would agree, however, that altering the data is not appropriate or ethical. This case is a good place to discuss the relative seriousness of various courses of action.

Suggested additional questions for discussion of Case 3:

1. What if Donovan had omitted the absorbance readings for those two points from her data set and from the graph? Is this approach better, worse, or as just serious as altering them?

2. What if Donovan had entered them in the data set and plotted them on the graph, but then ignored these two odd points when drawing the standard curve?

3. What should the response of the instructor be if Donovan's actions come to his attention? (This question could be posed for only one of the alternative actions or for each.)

Case 4

This case, with Professor East, is intended to be a direct parallel to Case 3 with Donna Donovan. The idea is to start the class thinking and discussing whether the context, class lab report versus scientific publication, can or should affect the ethical evaluation of a case. Here Professor East may be right that the test of compound 84 was flawed somehow, and if so, the timely publication of his theory will save chemical companies time and money in their efforts to develop new products. On the other hand, it is possible that his theory is incomplete and the results with compound 84 point to a flaw in the theory. Publication of the paper now may result in wasted effort and possible herbicides missed until someone eventually publishes a better theory.

There may be some discussion as to whether this is really an example of data alteration or rather professional judgement in the interpretation of the data set as a whole. In this case, as in many others, it should be pointed out that calling what East has done "interpretation" of data is really a rationalization for inappropriate data manipulation. However, as may come up here or in Case 8, drawing the line between interpretation and alteration is not always simple.

The same alternative actions could be proposed for Professor East as for Donna Donovan in Case 3:

1. What if Professor East had entered all of the data in the table and then noted that his new theory may not be able to predict all results? (Note that such an admission would most probably result in the reviewers denying publication.)

2. What should the response of the scientific community be if Professor East's actions become known?

3. Of what relevance, if any, is the difficulty, including time and expense, involved in resynthesizing compound 84? For example, would it make a difference if it took six months and $2,000, as opposed to one week and $20?

Case 5

This case is the first of three dealing with the omission of data. Omission is frequently considered to be much less serious than alteration, but should it be?

Case 5 is a variation on the scenario described in Case 4. Here, East is deliberately omitting data that do not agree with his hypothesis. He is omitting these data just so that he has the opportunity to present his theory, which he now has reason to question. The decision to omit the results for compound 84 should be considered to be misrepresentation and an unethical practice. In today's quest for publications, however, it is probably not a rare occurrence.

Some questions for further discussion of Case 5:

1. Is the data omission in this case better, worse, or just as serious a breach of scientific research ethics as the data alteration in Case 4?

2. What should the response of the scientific community be if Professor East's omission becomes known? Is your answer different than the one you gave for Case 4? If so, why?

Case 6

In this case, although Jamison did not choose the best course of action, that of reporting all of her results, one would have a difficult time justifying reporting her to the campus student ethics committee if her omission should become known. Here, there is no significant consequence if the data are reported or if they are not. Yet, in a professional setting, omissions could result in serious consequences (see Cases 5 and 7).

Another factor to consider is the reason Jamison decides to omit the saccharine result: concern that the instructor will treat her unfairly. A discussion of the teaching of science laboratories is appropriate here.

Should the ethics of an action be evaluated based on its consequences? Many would argue that it should be. The omission of data in a class lab setting is of no consequence to the rest of the world *except* that the class lab is the very place where future professional scientists learn their trade and their research ethics. Attention to research ethics is, therefore, more than just a nicety in these situations. There is a philosophical ethical theory that uses this "consequentialist" line of reasoning. Other ethical theories are based on consideration of rights or on virtues. (See chapter 2 of this volume.) The analysis of this or other cases through the application of various ethical theories might be a good topic for class discussion.

Some questions for further discussion of Case 6:

1. Are Jill Jamison's actions in omitting some data more like the actions of Able and Baker in Case 1 or Chase in Case 2? Is the change from alteration in Cases 1 and 2 to omission in this case a significant factor in your evaluation of the ethics of Jamison's actions?

2. Should the ethics of an action be evaluated based on the consequences of that action?

3. How does this case differ from Case 5? Are the differences significant?

Case 7

Here, as in Case 5, we have an omission of data issue in the professional setting—not malicious omission, but omission that borders on an unsound interpretation and that could have serious consequences for human health.

Some further questions to consider when discussing Case 7:

Assume that it is now two years later. You know everything that Dr. Kenworth did in the presentation of her data, and you are reevaluating her actions. What would your conclusions be in one or more of the scenarios proposed below?

a. Suppose that the subsequent clinical trials showed derivative 672 to be significantly more effective than Kayfab, in addition to being less expensive to manufacture and easier to store. Do these facts affect your ethical evaluation of Dr. Kenworth's actions? Consider that many additional lives were saved with this better antibiotic by not waiting a year to repeat the laboratory tests.

b. Suppose that the subsequent clinical trials showed that derivative 672 was much less effective than Kayfab. (The clinical results were all like those few "bad points" of the laboratory tests.) Do these facts affect your ethical evaluation of Dr. Kenworth's actions? Consider that some people participating in the clinical trials died when they were treated with derivative 672 instead of the proven Kayfab.

c. Suppose that as a result of further investigation of the "bad points," Dr. Kenworth detected a flaw in the test procedure and obtained new data that showed that derivative 672 was not as effective as Kayfab. As a follow-up to the experiment, she went on to modify Kenworth's Law so that it is now an even better predictor of antibiotic efficacy.

Case 8

This case may be a more difficult call for the students because the numbers weren't really changed. Instead, the data were represented in a different manner with "-" indicating both low and no detectable amounts of RNA. Assuming that the authors did not define the meaning of the symbol "-", one can argue that this is, at worst, an example of misrepresentation, not alteration.

On the other hand, the students may be more willing to call Hansen's and Ingalls's actions unethical first, because they are professional scientists, and second, because their published interpretation of the data differs from Grant's and they have ignored his objections. Ethical treatment of a colleague or student is a separate issue from the ethical reporting of scientific research, but an important one that you might wish to discuss. In your discussion of this case, an effort should be made to separate these two issues. After all, unpleasant people can be scrupulous in their publication of scientific results, while charming people can falsify data.

In discussing the second question posed following the case, the students should realize that they would want to know what the background level was for Grant's assays. They would want to know how many replicates were done of each plant, and if statistical treatment of the data showed the low amounts of RNA to be significantly different from background. Scientists can and should evaluate their data more objectively than was done in this case. With this information, it would be possible to determine whether or not the low amounts of RNA were essentially zero. Unfortunately, you sometimes don't have and can't get more information; but it is important for students to realize that often it is best to obtain more information before acting. Some of your students may also note that all the controversy is over Grant's data. What about Ingalls's data? One of the questions below addresses this point.

Some questions for further discussion:

1. What could or should Grant have done after Hansen and Ingalls dismissed his interpretation?

2. Was it appropriate to put Grant's name on the published paper as a coauthor if he disagreed with the interpretation of the experimental results and the conclusions drawn?

3. Note that Ingalls never produced his original data in this scenario. What if, at some point after publication, Grant's friend Lee, who is a graduate student in Professor Ingalls's lab, happens to run across Ingalls's original data

for the *lac* protein determinations. She discovers that Ingalls's results look just like Grant's: no protein detectable in two plants, very low amounts in 15 plants and high amounts in three. What should she do? Does this information change your original evaluation of this case?

You may wish to use this case to discuss the procedures that are to be followed in the reporting and investigation of possible research misconduct at your institution.

7. The Ethics of Genetic Screening and Testing

These case studies are meant to serve as an adjunct to a lecture on RFLP analysis or pedigree analysis in an undergraduate class in genetics or molecular biology. Several textbooks include descriptions of RFLP analysis (see, e.g., Weaver and Hedrick 1989: 309–313 for a discussion of the use of RFLPs in tracing the causes of Huntington disease).

The cases have been written for upper level undergraduates, that is, juniors and seniors. These materials can be treated as an integral part of the classroom lecture or they can be used in a separate discussion section run by teaching assistants. Students should come to class or discussion section prepared to discuss their answers and the reasons that underlie them. The instructor may want to ask students to imagine that they are genetic researchers who have been asked to answer the questions that follow each case.

We recognize that there may be no consensus on the answers to some of the questions. The cases have been written to stimulate discussion of important issues and increase students' awareness of the ethical dimensions of advances in scientific knowledge. Specifically, the cases have been designed to demonstrate some of the ethical implications of genetic screening, particularly the difficulty of weighing the potential benefits to humanity from increased knowledge of our genetic makeup against the risk of violating individual privacy rights and autonomy, as well as the pain and suffering that can accompany such increased knowledge.

Discussion of many of the issues raised in these cases, as well as other issues raised by genetic research, is provided in chapter 5 of the National Institutes of Health's *IRB Guidebook* (U.S. Department of Health and Human Services 1993: 5–42 to 5–63).

Case 1

Question 1. The most significant issues students should recognize in this case are those of privacy and informed consent, both of which are explored in subsequent questions. A more minor consideration is that it may be beneficial to know whether the girls are carriers so that if they develop anemia later in life they and their physician will know whether taking iron supplements will be an effective treatment. An anemia due to a lower amount of beta-globin is not responsive to increased iron in the diet, and excess iron can have deleterious effects.

Question 2. Most experts agree that voluntary testing and genetic counseling should be available to all potential carriers of beta-thalassemia. But this consensus leads to the issue of the age at which such testing should be provided. The instructor may also want to raise the issue here of whether it is appropriate to test persons who are otherwise unable to give informed consent to testing due to mental incapacity (including incapacity due to young age), cognitive impairment, or emotional instability. This might also be the time to discuss the need most people have for psychological counseling after they have received genetic counseling. They usually need help dealing with their feelings concerning the results and their decisions, regardless of whether the results are "good" or "bad." Do people have a "right" to such counseling? Do health professionals have a responsibility to provide it?

Question 3. Question 3 raises the issue of whether individuals should have the right *not* to be told certain information about themselves that they may prefer not to know. The question suggests further discussion around issues such as whether people should be forced to learn genetic information of other kinds about themselves, for example, results of tests for mental illness, propensity for alcoholism, nonapparent birth defects, late-onset diseases, and so forth.

A further question to pose here is whether persons who do not wish to have children should be forced to undergo mandatory testing. A strong argument can be made that persons who do not intend to have children should not be required to be tested, although the interests of employers and insurance companies would most likely argue the contrary.

Question 4. The issue in Question 4 is the relevance of the treatability of the genetic disease. PKU is treatable, whereas beta-thalassemia, like many genetic diseases, is not. With the availability of fetal testing for many untreatable genetic diseases, carriers of these diseases who wish to have children are faced with difficult choices. Some couples will choose to conceive and bear children regardless of the known risks, and even in the face of the knowledge that the child will have the disease. Other couples, expressing the view that they would not knowingly bear a child with a serious genetic defect, will choose either to adopt children or to abort any fetus diagnosed as having the disease.

Question 5. One aspect of the economic issue concerns the role of insurance companies in paying for genetic tests. Insurance subsidization of such tests is certainly the only way in which many individuals can afford them. How far should the definition of "potential carriers" be extended for insurance purposes? For instance, should a person whose cousin has CF expect his or her insurance company to pay if he or she wants to be tested? What evidence of the potential for genetic defect should there be before coverage becomes available?

There is an easily calculated statistical probability for any genetic relationship of the likelihood of carrying the same allele. For example, for the sibling of a carrier, the probability is 50 percent; for a nephew of a carrier it is 33 percent.

An additional question on this topic relates to the fact that insurance companies frequently will not pay for elective tests. The instructor may want to ask whether insurers should therefore be legally required to pay for genetic analysis of potential carriers.

Question 6. This question raises the general issue of the appropriate balance between confidentiality and the "right to know." The question implies several related issues that the instructor may want to raise, time permitting:

a. On the access issue, should the individual have a right to privacy of his or her medical records? What if the individual is a small child?

b. Regarding the results of testing, some diseases, such as manic depression and inherited colon cancers, may only develop later in life. Many of these diseases have only a statistical probability of developing. Homozygotes for inherited colon cancer, for example, may have a 30 percent chance of developing the disease in their lifetimes. What difference might these considerations make?

c. Insurance companies and employers have discriminated against carriers based on misunderstandings of the meaning of heterozygosity (i.e., the false belief that heterozygotes have or will develop the disease) or of the possibility that the carrier will later have a homozygous child who will require extensive medical expense. How can carriers be protected against such discrimination?

Do you think insurance companies would distinguish between carriers and persons who actually have a particular disease? If not, does this fact affect your answer to Question 5?

d. Does an individual's insurance company or employer have a right to know the results of such testing?

If the insurance company pays for such tests, is it necessarily entitled to receive the results? Why or why not?

If the insurance company pays for the tests, is it entitled to pass along the test results to the individual's employer (that is, when the policy is provided by or purchased through the individual's employer)?

e. Do persons or entities who might be affected by the results of a genetic screening test (e.g., employers and insurers) have legitimate rights that need to be protected? If so, how can their interests be fairly safeguarded? Is it necessary to sacrifice one interest in order to protect another?

f. Should other members of the individual's family be entitled to the results of such testing?

Should they be told of results that may affect them even if they don't request the information?

Is the nature of the disease relevant to your answer?

Questions 7 and 8. These questions are intended to create an awareness of the potential for misdiagnosis and the ethical problems that surround such potential. These questions present a good opportunity to discuss the role of the genetic counselor and the importance of proper training in such counseling. For example, the possibility of misdiagnosis means that genetic counselors must be very careful when informing patients about the results of tests, indicating what they are based upon, the statistical probability of accuracy, and so forth. They also need to point out clearly the distinction between being diagnosed as a carrier for a particular disease and being diagnosed as having that disease, and the implications of each.

The potential for error also counsels in favor of maintaining the confidentiality of test results, as insurance companies and employers may very well discriminate against individuals based on genetic test results that later turn out to be erroneous.

At a more general level, these questions can also be used to discuss the tension researchers face between the felt need to withhold research results until the social and policy implications of the findings have been considered versus the desire to disclose scientific knowledge to the public.

Case 2

Question 1. In an actual case involving similar facts, the insurance company paid for the genetic screening, then refused to pay for further tests or treatment once the mother decided not to have an abortion. It eventually reversed its decision and agreed to cover pre- and postnatal costs.

Question 2. These questions bring up ethical issues of the value of human life, which are dealt with at length in Veatch (1989). One factor that will inevitably come up in the discussion of this case is the severity of the disease or disorder and the availability of treatment. The follow-up question can be used to push at the edges of the issue. What if the disease is completely treatable, if not curable? What if it is treatable to some extent, but with serious difficulties remaining? Retinoblastoma is a disorder of the latter kind; polydactyly (excess fingers or toes) is an example of the former.

A bit of background information on retinoblastoma may be helpful. Retinoblastoma is inherited as a dominant trait. Ninety percent of the people who inherit the defective gene develop retinal tumors by the time they are four years old. If monitored and treated early, the tumors can usually be prevented from causing blindness or death. Before monitoring was available, retinoblastoma was a fatal disease. Still, if not treated early enough, the eye might have to be removed. And, even if the retinal tumors are discovered and treated early, persons with retinoblastoma experience a higher incidence of cancers later in life, particularly bone cancers.

Question 3. Although lawsuits for "wrongful birth" are relatively recent, as of this writing, hundreds have been filed. Courts have generally allowed the parents to recover against the physician for misdiagnosis. For example, in *Becker v. Schwartz*,[1] the New York Supreme Court held that a doctor was liable for giving parents of a child born with polycystic kidney disease inaccurate information to the effect that there was almost zero risk that the child would have the same genetic disease that their first child had.

Recovery is sometimes limited to the costs relating to the child's disability caused by the genetic defect that the physician failed to diagnose (see Dworkin 1986: 96). In some cases, however, courts have allowed the parents to recover the entire costs of raising the child.[2] In a couple of states, in keeping with restrictive abortion laws, legislation prohibits recovery by the parent if the testing that should have resulted in a diagnosis of the potential defect or disease is done at any time after conception.

Courts have been reluctant to recognize "wrongful life" cases brought by parents or guardians on behalf of children born with genetic defects that their physicians were negligent in failing to diagnose. See, e.g., *Goldberg v. Ruskin*, 499 N.E.2d 406 (Ill. 1986) (rejecting the wrongful life claim of a child born with Tay-Sachs disease); *Bruggeman v. Schimke*, 718 P.2d 635 (Kan. 1986) (rejecting a wrongful life claim brought by a three-year-old boy claiming that he would not have been born except for the negligence of the defendants in giving his parents improper genetic counseling). A few courts, however, have recognized such claims. See, for example, *Continental Casualty Co. v. Empire Casualty Co.*, 713 P.2d 384 (Colo. App. 1985), and *Procanik by Procanik v. Cillo*, 97 N.J. 339, 478 A.2d 755 (1984).

Question 4. Although it does not raise a genetic screening or research ethics issue per se, this question is included to provoke discussion on the appropriate role of law in relation to private reproductive matters and the ethical value of human life of persons born with genetic defects.

Case 3

Question 1. Potential problems with the Power Force Company's genetic screening program include violations of employee rights to privacy and confidentiality and the employer's abuse of information. The employer's abuse would be evidenced through its use of test results to discriminate unfairly against employees.

Question 2. Screening current employees is ethically more problematic than screening prospective ones because of the lack both of advance notice and of choice of participating in the testing as a condition of employment.

Question 3. It is important to point out the risk that employers will use genetic screening to exclude susceptible employees from exposure to hazardous toxins in lieu of taking efforts to make the workplace safer for all employees. Also, in discussing option (c), students should recognize that jobs in other areas of the company may have a different (i.e., lower) pay scale and benefits package, be less pleasant, or be in a distant location.

Question 5. This question is intended to address the issue of exclusion of groups rather than individuals. What if the company adopted a policy prohibiting groups in which the hypersusceptibility trait appears from working with lead to avoid the expense of testing individuals?

Question 6. The legal case entitled *International Union, U.A.W., et al. v. Johnson Controls*[3] involved an employer's attempt to impose a so-called fetal protection policy prohibiting all women of child bearing age from working for the company. The company attempted to justify its decision by explaining that working for the company involved exposure to lead (which is dangerous to fetal development), and that such exposure could not be eliminated by removing lead from the workplace because no alternative to its use is available. In March 1991, the U.S. Supreme Court held that the company's fetal protection policy unconstitutionally discriminated on the basis of sex by violating women's right to equal protection of the law.

Case 4

Question 1. This case raises issues surrounding the testing of minors, particularly given the questionability of whether informed consent can be obtained from minors below a certain age or stage of development. There is a split of opinion among medical ethicists concerning whether it is morally permissible to test minors for any purpose that is not therapeutic, that is, not intended for the minor's own benefit. Some argue that such testing is permissible if it involves minimal risk of harm to the child (see, e.g., McCormick 1981: 51–57 and 98–113; and Veatch 1987). Others argue that such testing is impermissible regardless of the risk of harm or the existence of "proxy" consent by the parents (see, e.g., Ramsey 1970: 1–58 and Ramsey 1976: 21–39). Where the testing is being done for research purposes, federal human subjects regulations, which place specific restrictions on research involving children and minors, will apply.[4]

Question 2. The case in favor of testing is arguably stronger when the results may have an impact on the individual's immediate future than when they are not likely to influence the individual's apparent health and well-being for an unknown (and significant) number of years. In addition, testing is arguably more justifiable when the link between having the gene and developing the disease is direct and inevitable (although there may still be a question of *when* the disease is likely to develop, as is currently the case with HD), than when the linkage is indirect.

Question 3. Mandatory testing is arguably more justifiable when there is a cure for the disease being tested, since positive test results can be dealt with constructively, in contrast to situations where no cure exists, so that the knowledge gained from test results is likely to cause significant pain and suffering. One can also argue, however, that the knowledge of positive test results allows one to plan appropriately for the future.

Question 4. Even if testing and diagnosis might not lead to a cure for the particular individual being tested, it is arguably more justified where it has the potential to lead to a cure for others than when there is no direct connection between the testing and research.

Question 5. This question again raises the issue (discussed in connection with Case 1) of whether, as a society, we should protect employers' (and others') interest in knowing the results of genetic tests performed on individuals.

You may want to ask this follow-up question: Suppose Jay is 17 rather than 20 and is applying to West Point Academy (where the U.S. taxpayers will pay to train and educate him) as the first step in an anticipated military career. Should he be required to disclose his family genetic history and undergo testing as part of his admission requirements?

Question 6. For further information on the sequence of the HD gene and on speculation concerning the repeat array length refer to The Huntington's Disease Collaborative Research Group (1993) and Angier (1993).

8. Ethics and Eugenics

This chapter has been designed for use with undergraduate students in connection with instruction on pedigree analysis in an introductory biology class or in a genetics course for majors. The pedigrees were chosen to be amusing as well as to demonstrate pedigree analysis. They illustrate the kinds of research that made the eugenics movement one of the classic chapters in the "nature versus nurture" controversy.

For further information on the link between eugenics and modern technologies in human genetics you may wish to consult chapters 1, 13, and 14 of Kevles and Hood's *The Code of Codes: Scientific and Social Issues in the Human Genome Project* (1992).

Some considerations relating to the cases as well as discussion questions that you may want to raise with your class are outlined below.

Case 1

Question 1. The pedigree indicates that the trait for naval heroism could be transmitted as an autosomal sex-limited dominant (IV1 inherited the allele from his mother who could not express the trait) or as a Y-linked trait with reduced penetrance (note II2, III1, III10, and V5).

The following are additional questions for discussion you may want to ask:

1. If women had been allowed to serve in the navy, would this allele still be transmitted as sex-limited?

2. At one time, pedigrees like this one appeared in all science textbooks. Now that intelligence and other personality traits are no longer considered to be linked exclusively to genetics, such pedigrees are no longer used. Who or what is responsible for this change?

Question 2. This question is meant to induce discussion on the components of a behavior. It is worth making the point that although the eugenicists' claims were excessive, heredity does matter. There are differences between individuals in talent and intelligence, and much of that is inborn.

Question 4. This question points out a couple of ways in which positive eugenics is practiced today and may be expanded in the near future. One can argue that some existing dating and introduction services are de facto devices for positive eugenics as are exclusive schools and organizations.

Question 5. New technologies in human genetics have recently given us the opportunity to select which, of a number of in vitro fertilized embryos, will be used for implantation. What criteria should parents be allowed to use to decide which embryos will be selected? Freedom from a deleterious allele associated with a genetic disease? Inheritance of desirable alleles? Gender?

In the not too distant future, we will probably be able to introduce new alleles. Should this be allowed? If so, who decides what types of genetic modifications will be available?

You may wish to consult Bishop (1993) or Handyside et al. (1992) for information on preimplantation selection.

Case 2

Question 1. Feeblemindedness appears to be transmitted as an autosomal recessive trait.

Question 2. At first glance, criminality also appears to be inherited as an autosomal recessive trait, particularly if only Pedigree 2 is considered. However, if one also considers Pedigree 1, one discovers that many of those who would be heterozygous carriers of the feeblemindedness allele, *f*, are criminalistic. The information on this family is consistent with a hypothesis that criminality is an incompletely penetrant dominant trait of the feeblemindedness allele. Thus, if an individual is homozygous *ff*, he or she is feebleminded. If he or she is heterozygous *Ff*, he or she has an approximately 40 percent chance of criminality. This is summarized in Pedigree 4.

You may wish to pursue this further discussion question: Is it likely that there were no alcoholics, adulterers, or criminalistic persons in the naval hero pedigree, or is it more likely that the researchers were not looking for these traits?

Question 3. The general purpose of this question is to motivate students to recognize that many factors besides genetics may influence behavior or mental capacity.

Questions 4–6. These questions are intended to raise students' awareness

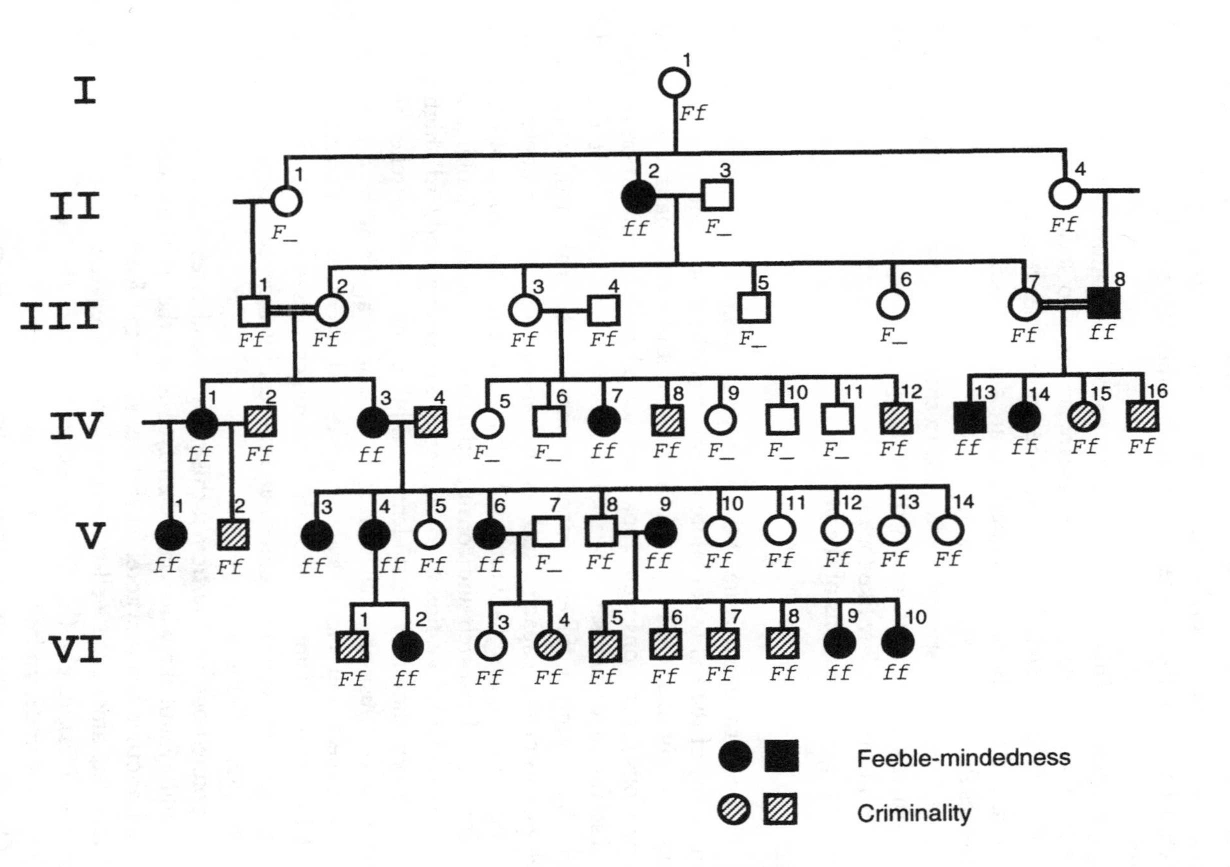

Pedigree 4. Summary: Feeblemindedness and criminality.

of the ethical dimensions of research, particularly research concerning genetic traits. One issue that should be brought out is the level of certainty researchers should have before reporting their findings on matters about which there is much debate and disagreement, such as the inheritance of the traits discussed here.

Question 5. This question should encourage students to think more broadly about the responsibilities researchers have to society relating to the results of their research. This includes responsibilities regarding the topics of research and how the research questions are designed and implemented as well as the results of the research itself.

Question 6. Question 6 is intended to spark a discussion of the potential for negative eugenic applications of emerging technologies in human genetics and reproduction. A law like the one given as an example would be analogous to the forced sterilizations of the past.

Case 3

This case raises the issue of the appropriate limits of the interpretation of data. Many of the Kallikak children were given the Binet-Simon test (referred to at the beginning of Case 2) to generate objective scores, but the adults and the previous generations were apparently scored as feebleminded on the basis of observation or hearsay alone. Students should be aware that the field-worker's assessments of feeblemindedness may have been affected by prejudices irrelevant to mental capacity, such as an aversion to poverty or uncleanliness. *Questions 1–4, 6, and 7* are intended to raise students' awareness of the social ramifications of research and consideration of the researcher's obligations to the larger society. *Question 6*, for example, concerning Goddard's high grade moron theory, should cause students to think about how researcher bias can change the conclusions reached from a piece of data. We all tend to observe what we expect, and scientists are no different from others in this regard.

A further discussion question that you might wish to explore: If the dividing line between feeblemindedness and normal intelligence is this subtle, is it an appropriate division?

As pointed out by *Question 5*, this pedigree indicates that no one among the "good" Kallikaks who was heterozygous for the feeblemindedness gene married another heterozygote. Therefore, either the feeblemindedness allele was very rare among the well-to-do (i.e., it was segregating quite rigidly along economic lines), or else the different environments of the two families was a major factor in determining the fates of the children.

Question 8 is intended to trigger a discussion concerning how we determine whether a trait is neutral, desirable, or undesirable. With the advent of new medical treatments and molecular biology there may be a tendency to reclassify less desirable traits, such as short stature, as genetic diseases to be treated or, if possible, avoided (see Bishop 1993). Nancy Wexler has also written concerning her fears that those having what are considered to be undesirable alleles will be considered inferior and will be discriminated against (Wexler 1993).

9. Ethical Issues in Animal Experimentation

Guidelines for the conduct and procedures of IACUCs are provided in the federal regulations.[5] You may wish to consult these guidelines in order to add "realism" to the role play exercise. If your instructional time for these cases is limited, you may want to forgo the role play format of having students play the part of IACUC members, and simply use the Questions for Discussion, or those specific questions that are most relevant to the material you are covering in your class.

Additional material on the philosophical positions underlying the animal rights movement and suggested exercises for classroom discussion, including cases, is contained in Herzog (1990). In addition, the United States Department of Agriculture has produced a bibliography containing 270 citations on the topic of "Ethical and Moral Issues Relating to Animals" (Clingerman 1988).

Students will need instructional guidance in distinguishing between the relevance of the *legality* of using animals versus that of *ethical* or *moral* justifications for either using or not using animals in research. A useful forum debating the question of civil disobedience as a means of halting the use of animals in experimentation is provided in Donnelly and Nolan (1990).

Instructors may be interested in knowing that a survey of 263 students at State University of New York at Albany in 1987 revealed that although most students are concerned about pain and suffering of animals (76 percent), they prefer to allow research to be conducted using animals rather than to allow humans to suffer and die from disease (85 percent), and also prefer to have new forms of surgery and experimental drugs tested on animals before permitting their use on humans (67 percent). The majority of students also felt that animal experimentation was necessary (62 percent). The authors of the study conclude by recommending the inclusion in discussions with students of information about government regulations and professional codes relating to the humane care and treatment of animals used in research (Gallup and Beckstead 1988: 476).

Case 1

It may be useful to provide students with some background information on institutional review committees—how, for example, they came into being and how they work. Students should be pushed to question how decisions on approving or disapproving research with animals are made. Discussing the institutional review committee is a good vehicle for drawing out consideration of how to go about making ethical decisions—distinguishing, for instance, between facts and values or opinions, and between the value of expert evaluations and the evaluations of concerned citizens.

Although Questions 1(a) and (c) may seem to ask the same question, they are intended to raise two very different issues. Question 1(a) asks students to think about how they will decide whether a given project has scientific merit. Question 1(c) asks them to decide whether (and to what extent) the scientific importance of research is even relevant to deciding whether the proposed use of animals is ethically acceptable. An advocate of animal rights would contend that the utilitarian weighing of harms against benefits is irrelevant: no matter how important the research is to the elimination of human suffering, no animal life can be sacrificed for the benefit of humans.

The following are some suggested follow-up questions to those presented in the Questions for Discussion that you may want to use in your class, time permitting.

PART A.

1. If students would allow the experiment, ask whether there should be limits on the number of rats that can be selectively bred for use in experiments such as these and, if so, what they should be.

2. If students would not allow the experiment, ask whether *any* research would be important enough to be allowed and, if so, what kind of research.

PART B.

Ask the students more about SKN's conduct:

1. If you think SKN's conduct was unethical, is your objection based on form (the means used to carry out the protest) or content (the ethical justifiability of the protest), or both?

a. If you think the *form* of the protest is unethical, what methods would you recommend to SKN to make their activities less objectionable?

b. If you think the *content* of the protest is inappropriate, what circumstances do you think would make SKN's protest legitimate?

2. What steps, if any, would you recommend that the university take to sanction the people involved in SKN's activities?

Case 2

The following are issues raised by Dr. X's research that you may wish students to take into account as they decide what their IACUC should do.

1. To what extent does the theoretical or practical importance of Dr. X's

research justify the procedures performed on the monkeys—deafferentation of the monkeys' arms and recording from the cortex followed by euthanasia?

2. Under what circumstances would killing these animals (euthanasia) be justified? Note that the animals cannot be put in a zoo with other monkeys, as they would be harassed as cripples and would risk being harmed or even killed. If the animals are not euthanized, who should pay for their care?

3. Should the ethics of experimentation on animals depend on the amount of animal suffering involved? If so, how does one measure or evaluate this suffering?

4. From an ethical standpoint, is there a difference between causing pain to animals during the research and euthanizing them at the end of the research, assuming that both are necessary to complete otherwise valid research?

5. Suppose that Dr. X was able to demonstrate that he had complied with all relevant government regulations governing the treatment of experimental animals. Does compliance absolve him from charges of unethical treatment of animals?

6. Did Ms. Y and LAN choose an appropriate strategy for exposing Dr. X? What are ethically appropriate and inappropriate ways for animal rights advocates to oppose animal experimentation?

7. Should the federal funding have been withdrawn from Dr. X's laboratory before the court case was decided? What other punishment or reprimand, if any, do you think would have been an appropriate sanction?

10. Research Involving Human Subjects

This experimental "proposal" was developed from a published study (Pihl et al. 1984). Students should be able to identify the elements of deception: the use of a placebo beverage (attempting to deceive subjects into believing that they are drinking alcohol when they are actually consuming a nonalcoholic drink), and leading subjects to believe that they are interacting with another person in a pain-perception/reaction-time task, when, in fact, they are interacting with a computer program. In addition, subjects are being led to believe that they are causing another person pain. You may wish to read the published study prior to discussing it with your class. They may be interested to learn that the experimental results showed significantly less aggression in males who consumed (or thought they had consumed) beer than those who consumed (or thought they had consumed) distilled spirits, and nonsignificant differences between subjects who received alcohol and those who received the placebo.

Students should read the case before the class in which it is to be discussed. You may ask them to answer the questions individually before class and then hold a class discussion on why individuals reached the decisions they did. Alternatively, you might ask students to meet in small groups for the first 10–15 minutes of class to work through the questions together and then reconvene the class for a group discussion.

You may wish to structure the class discussion to follow the format of an IRB committee meeting convened to decide whether the alcohol experiments should be permitted to proceed, based on the information contained in the proposal. You may wish to play the role of committee chairperson or monitor, or ask a student to volunteer for the role. Detailed descriptions of the procedures followed by IRBs are contained in U.S. Department of Health and Human Services (1990), Veatch (1987), and Beauchamp et al. (1982). If time is limited, you may simply want to ask individuals to volunteer responses to the questions you consider most pertinent.

The following are suggested questions for further discussion:

1. If time permits, a class discussion could be developed around the following general issues raised by nearly all experiments using human subjects:

   a. Cost/benefit. What general factors determine "acceptable risk" for subjects in an experiment? Would extreme stress ever be justified?

   b. Informed consent. What is it? Are exceptions or substitutions ever justified? Does signing a consent form always constitute "informed consent"?

   c. Stress. How do you determine an acceptable level of "stress"? Is "psychological stress" less serious than "physical stress"?

   d. Long-term harm. How could long-term harm be assessed?

   e. Deception. Is deception of subjects a necessary characteristic of all psychological experiments with human subjects? Is the use of deception unique to psychological experiments? (Consider the use of placebos in medicine.)

2. Ask the class to draft an informed consent form that would avoid the deception employed in the alcohol experiments, yet enable the investigators to obtain data on alcohol-related aggression. This exercise might involve small groups or individual take-home or in-class assignments, or an at-large class effort using the blackboard to record the results of class "brainstorming."

3. The Milgram experiments,[6] in which subjects were deceived into thinking they were administering harmful shocks to participants involved in a learning experiment, are often cited as an example of problematic treatment of subjects. Is the study described in the present proposal more acceptable than the Milgram studies? Why or why not?

4. The American Psychological Association (APA) has formulated guidelines for research involving human subjects. Principle C states: "The investigator always retains the responsibility for ensuring ethical practice in research. The investigator is also responsible for the ethical treatment of research par-

ticipants by collaborators, assistants, students, and employees, all of whom, however, incur similar obligations" (APA 1987: 31).

a. Is this Principle appropriate?
b. Does the experiment presented in this case comply with Principle C?

5. APA Principle D states: "Except in minimal-risk research,[7] the investigator establishes a clear and fair agreement with research participants, prior to their participation, that clarifies the obligations and responsibilities of each. The investigator has the obligation to honor all promises and commitments included in that agreement. The investigator informs the participants of all aspects of the research that might reasonably be expected to influence willingness to participate and explains all other aspects of the research about which the participants inquire. Failure to make full disclosure prior to obtaining informed consent requires additional safeguards to protect the welfare and dignity of the research participants. Research with children or with participants who have impairments that would limit understanding and/or communication requires special safeguard procedures" (APA 1987: 31–32).

a. Has the experiment presented in this case complied with the requirements of Principle D?
b. If the requirement of full disclosure has not been met, have the additional safeguards mentioned been adopted?
c. If compliance with Principle D is impossible because providing the required information would preclude doing the experiments, should the experiments be allowed to proceed?

6. Should the APA Guidelines be viewed as definitive with respect to the ethical questions raised in experimentation using human subjects? What is the role of professional codes in relation to research ethics?

11. Research Involving Human Subjects: The Administration of Alcohol

The primary issue presented by these cases is whether administering alcohol to subjects represents a potential harm to their well-being, or, in which contexts and under which conditions might alcohol administration be harmful (e.g., pregnant women, people at risk of alcohol dependency, people with a history of abuse, alcoholics, using incentives that might encourage subjects to be dishonest). A good discussion of the issues raised by the participation of human subjects in drug and alcohol research appears in the National Institutes of Health's *Institutional Review Board Guidebook* (U.S. Department of Health and Human Services 1993: 5–64 to 5–70). In addition, the National Advisory Council on Alcohol Abuse and Alcoholism at the National Institute on Alcohol Abuse and Alcoholism has issued *Recommended Council Guidelines on Ethyl Alcohol Administration in Human Experimentation* (1989), which you may wish to consult when working with these materials. In addition to discussing the involvement of alcohol dependent individuals in research (see notes regarding Recruitment Scenario 4), the Guidelines address such issues as obtaining family histories; the subject's age; medical and psychological evaluations; the involvement of women of childbearing age; medical backup services; alcohol-naive individuals; administering to occasional drinkers larger amounts of alcohol than those to which they are accustomed; deception in alcohol research, including the use of placebos; and payments to subjects.

Recruitment Scenario 1

Question 2. It is not uncommon to pay subjects for participating in research; such payments are not prohibited by the federal human subjects regulations. Paying participants, however, raises the question of the appropriateness of the payment: How much is enough? How much is too much? At bottom is the possibility that the payment will constitute an undue inducement to participate. Payments must therefore be commensurate with the risks, inconvenience, and investment of time that subjects are being asked to undertake. In setting the type and amount of incentive, investigators (and IRBs) need to consider whether the incentive is likely to inhibit subjects' ability to make a reasoned decision whether to participate, and also whether it may induce them to conceal important screening information in order not to disqualify themselves from participation.[8]

Recruitment Scenario 2

In addition to the issues raised by the first scenario, this recruitment method presents concerns of risk and confidentiality. You should explore with the students the risks associated with administering alcohol to persons with either a personal or family history of alcohol or other substance abuse. You should also examine concepts of confidentiality, how confidential information can and should be protected, methods of collecting confidential information, and related issues. Issues of consent are also important here.

Recruitment Scenario 4

The administration of alcohol to alcohol dependent persons presents some difficult ethical issues. Not only is there the question of undue inducement (i.e., subjects may agree to participate in order to receive the substance to which they are addicted), but also concern about risk. A potential for toxicity exists, which must be carefully monitored by the investigator. Adequate medical supervision should be provided. A further consideration is conflict of interest where, for example, the investigator has a tie to the treatment clinic. Such ties may be financial or simply "good will." If the clinic has opened its doors to the investigator to recruit subjects, will referrals of subjects for treatment be impartial?

The National Advisory Council on Alcohol Abuse and Alcoholism's "Recommended Council Guidelines," which address the administration of alcohol to alcohol dependent persons, state:

> Experimentation which requires individuals who are alcohol-dependent or alcoholics to be exposed to alcohol clearly warrants special attention. There are a number of extremely important principles which need to be addressed by anyone considering or evaluating requests to undertake such research. It is noted that these issues differ to a degree, depending on where in the disease/rehabilitative/recovery process the potential subjects are. Further, it is useful to distinguish between these stages in addressing some of the key issues. For example, the likelihood that a subject would otherwise be encountering the agent (alcohol) would clearly differ, depending on their disease or recovery status. The risk of the investigator inflicting harm is clearly greater when the probability that the subject would be otherwise exposed to alcohol is lower.
>
> When potential subjects include alcoholics who are current, active drinkers, the screening procedure must clearly include a medical examination to assure the absence of any medical or mental condition for which further alcohol exposure at the dose contemplated would be contraindicated. Further, the Council stresses that it is incumbent on the investigator, or his/her agent, to make a serious and concerted effort to link such individuals with treatment. This linkage should be active in bringing together the subject with alcoholism treatment personnel, and not passive as in only providing names of treatment programs and phone numbers to the research subject. Whether or not the subject chooses to remain in the treatment program, it is incumbent on the investigator to actively facilitate entry of the research subject into the program.
>
> The use of subjects who have completed the initial phase of treatment and progressed into rehabilitation or recovery would require an extremely strong scientific justification *and* risk/benefit assessment. Different factors will need to be considered, including at what stage they are in the rehabilitation program and the alcohol dose employed. Both the research staff and the treatment personnel must consider the potential for untoward effects on the treatment/recovery process. There should be a continuation of treatment after conclusion of research participation for a sufficient period to ensure continued recovery.
>
> At the present time (1989), it is considered inappropriate to administer alcohol to any recovering alcoholic who is abstinent and living a sober life in the community. In taking this position, the Council believes that the issue of risk for relapse outweighs any consideration which may be afforded to the willingness of the subjects to participate in the project through informed and voluntary consent, or the unique requirements within a study to include recovered alcoholics to address the hypothesis posed. This position is derived

> from an assessment of risk, since the risk of the exposure eliciting relapse (or other health problems) is considered too great to warrant the recovering alcoholic's participation (1989: 5–6).

The guidelines set out the issues that the council feels should be addressed by investigators conducting research that involves the administration of alcohol.

12. The Ethics of Deception in Research

These studies were conducted in an effort to study the factors that influence whether or not bystanders will intervene in different social contexts. The research was inspired by questions arising from the murder of Kitty Genovese in New York City in the late 1960s, a murder that was witnessed in some manner by 35 bystanders who failed to intervene in time to help Ms. Genovese. In none of these experiments were the subjects aware of the controlling conditions that were the independent variables. In fact, in three of the experiments (Cases 1, 2, and 5), the subjects were not even aware they would be involved in an experiment. Each case raises ethical issues relating to the use of deception in psychological research, including: (1) subjects' informed consent; (2) debriefing of subjects; (3) violation of subject's rights; and (4) the ethical status of each case relative to the others.

After discussing the five cases with the class, the instructor might ask students to rank order them from least to most ethical, and provide reasons for their ordering. If time allows, the instructor may also want to engage the class in a discussion of some of the larger issues raised by these cases. The following are some suggestions:

1. What are appropriate methods for carrying out experiments to test bystander willingness to intervene?

2. Is there an ethical difference between observing people's behavior without intervention and manipulating their environment?

3. Is there an ethical difference between fully informing subjects about the subject of the research, soliciting volunteers to participate in "an experiment," and giving them no information at all to indicate that they are being involved in an experiment?

4. Does the public or private nature of the space in which the experiment is conducted bear on its ethical appropriateness?

13. Misconduct in Science

Part B. The issue of animal use can be further explored by posing the following questions:

1. If you observe someone in your laboratory decapitating rats without first anaesthetizing them (assuming that the protocol calls for anaesthesia), are

you obligated to report the activity? To whom would you report it (e.g., the lab head [assuming he or she is not the person in question], the animal care and use committee, the campus misconduct committee)?

2. Suppose J had included the unanaesthetized decapitation in the protocol, and the animal care and use committee refused to approve the protocol under those conditions. What should J do? What should Professor X do?

Parts C and D. The point of these questions is to explore the distinction between misleading reporting of data and giving unnecessary details. A further distinction, however, should be made between the level of detail that *must* be recorded in laboratory notebooks (i.e., as much detail as possible) and the level of detail that can and should be provided in a publication. It will be important to acknowledge the reality that the allowable length of manuscripts is limited, but also to make the point that the author must make an appropriate judgment about what must be included to report the research results accurately.

Part E. This question presents the opportunity to discuss the appropriate way to handle conference and publication submissions under conditions of extreme time pressure. Honesty in the reporter's phrasing of exactly what has been accomplished and what remains to be done (e.g., "Preliminary results indicate . . . ") are paramount.

14. Science and Coercion

Case 3

You may want to use this case to explore with the students the notion of a randomized clinical trial and the ethical issues RCTs raise. In the present case, for example, if cognitive therapy is recognized as the standard treatment for depression (or even as a "better" treatment in most cases), then the assignment of the control subjects to discussion groups, without some therapeutic justification, is problematic. Furthermore, the concept of "equipoise" in randomized trials (i.e., the notion that random assignment to treatment groups is ethical only if the investigator can honestly state a null hypothesis) might be examined (see Freedman 1987 and U.S. Department of Health and Human Services 1993).

Case 4

Question 3. This question tries to get at the problem of professional status and its concomitant responsibilities. A nonscientist in a nonuniversity setting might be free to make the inquiries that Professor X has made. He or she might even call it a "study." Yet there are no institutional oversight mechanisms for those inquiries (unless the research involves a regulated activity or one that is otherwise restrained by law or institutional policy). While the federal regulations governing the involvement of human subjects in research do not apply to nonfunded research, many institutions, as a matter of policy, require that all research involving human subjects be reviewed by the IRB; many apply the regulations or some relaxed version of the regulations to all nonfunded research. You might explore the conditions that give rise to this apparent discrepancy.

15. Behavior Control

Cases 1, 3, and 4 involve the use of behavior control techniques outside of the research setting. Nonetheless, the principle of respect for persons (autonomy) is relevant here, particularly where control is being applied as a form of medical or psychological treatment. An important issue, therefore, is consent to treatment and how it can appropriately be obtained.

Cases 2 and 5 involve research on behavior control techniques. You might wish to explore the application of the human subjects regulations to these cases. In particular, Case 5 raises the issue of the involvement of children in research. As background, consider the following:

The federal regulations require IRBs to classify research involving children into one of four categories:

1. Research not involving greater than minimal risk;
2. Research involving greater than minimal risk, but presenting the prospect of direct benefit to an individual subject;
3. Research involving greater than minimal risk with no prospect of direct benefit to individual subjects, but likely to yield generalizable knowledge about the subject's disorder or condition; or
4. Research that is not otherwise approvable, but which presents an opportunity to understand, prevent, or alleviate a serious problem affecting the health or welfare of children (45 *Code of Federal Regulations* 46.404–46.407).

Research in the first category may be approved by an IRB if adequate provision is made for obtaining the assent of the child and the permission of the parents. Research in the second category may be approved only if (a) the risk is justified by the anticipated benefit to the subject and (b) the relationship of risk to benefit is at least as favorable as any available alternative approach. Research in the third category may be approved only if (a) the risk represents a minor increase over minimal risk; (b) the intervention or procedure presents experiences to subjects that are reasonably commensurate with those inherent in their actual or expected medical, dental, psychological, social, or educational settings; and (c) the intervention or procedure is likely to yield generalizable knowledge about the subject's disorder or condition that is of vital importance for the understanding or amelioration of the subject's disorder or condition. Research in the fourth category may be approved only if the IRB, the Secretary of Health and Human Services, and a panel of experts finds that the research presents a reasonable opportunity to further the understanding, prevention, or alleviation of a significant problem affecting the health and welfare of children. In addition to any other requirements, the permission of the parents and, where appropriate, the assent of the child must be obtained. Circumstances in which the child's assent might not be required include cases in which the research offers the child the possibility of a direct benefit that is important to the health or well-being of the child and is available only in the context of the research (45 *Code of Federal Regulations* 46.404–46.408; see also U.S. Department of Health and Human Services 1993: 6–18 to 6–25).

Consider exploring the following questions regarding the involvement of children as research subjects:

1. Into which category of research does Case 5 fall?

2. Should Rich's assent be sought? If he declines to participate, must his wishes be honored? Of what relevance is Rich's age?

The following are some suggested questions and issues for class discussion on the cases taken as a whole.

1. Focusing on underlying ethical principles, how are the cases alike? Make an attempt to rank the cases from most ethically acceptable to most ethically unacceptable and justify your ranking.

2. The preceding cases were concerned with *reducing or eliminating* specific behaviors that the controller considers desirable—either by administration of aversive stimuli or by removal of positive rewards contingent upon occurrence of behavior. The student should realize that procedures designed to *facilitate* behavior that the controller considers desirable—either by administering positive stimuli (rewards) or removal of aversive stimuli contingent upon the behavior—may be no less problematic from an ethical point of view. As an example of control through administration of positive stimuli, consider the rigid control of behavior exerted in token economies in such places as psychiatric hospitals, where self-care activities (such as making one's bed, combing one's hair, or cleaning one's plate) are rewarded with tokens that can be exchanged for treats or privileges (such as watching TV, going for walks, or receiving visitors). As an example of control by removal of aversive stimuli, consider the practice in some totalitarian countries of incarcerating objectors in intolerable conditions until they enthusiastically endorse the party line.

3. The behaviors that are exhibited by the individuals in the cases in this chapter were determined to be "abnormal" by psychological researchers. As these cases demonstrate, labeling certain behaviors as abnormal is not inconsequential, but rather may have significant ramifications for persons who display such behaviors. For example, physicians, therapists, or counselors may try to control such persons' behavior using more drastic techniques than they would use with other individuals. Children may be more likely to be segregated into special classes in school; adults as well as children may be more likely to

be treated with medications or institutionalized. Regardless of the severity of the treatment, the fact is that once a behavior is labeled "abnormal," persons who exhibit such behavior are treated differently. Explore with the students the role that researchers play or should play in our society in labeling some forms of behavior as abnormal.

4. As a psychologist attempting to design ethically acceptable principles of behavioral control, what morally pertinent variables should be kept in mind?

16. The Historian's Code of Ethics

Before beginning discussion of these cases, make certain the students have a clear idea of what plagiarism is and some conception of why it is wrong. Students should be able to identify the principles underlying the AHA "Statement on Plagiarism," including the research community's commitment to and reliance on the principles of accuracy and honesty and the right to protection from misappropriation of private property. You might ask students whether either of these principles, alone, is an adequate basis for the prohibition on plagiarism, or whether they can suggest a more adequate rationale.

Case 2

The question of what responsibility a graduate student's dissertation supervisor has for the student's work is an important point for discussion. You might also ask students to discuss the role that various other actors should have in identifying and correcting instances of plagiarism and in disciplining perpetrators (e.g., the publisher, the referees, the authors whose works have been plagiarized, the AHA).

17. The Use and Interpretation of Historical Documents

This case is designed for use with undergraduate students who will be writing research papers. It could also be used to teach about problems of documentation in any kind of writing. Students should be able to identify the appropriate or inappropriate uses to which the text is put. You could assign the case as an out-of-class exercise that would then be discussed in class. Discussion could be done with the class as a whole or in small groups.

The following is an instructor's guide to the appropriateness of each statement in the "essay" following Wilberforce's speech.

Mr. Wilberforce and Mr. Norris held differing views on the African slave trade.[1]

[1]William Wilberforce, Speech before the House of Commons, *Parliamentary History*, 28, 45–47, as quoted in: *Cardinal Documents in British History*, R. L. Schuyler and C. C. Weston, comps. (Princeton, NJ: D. Van Nostrand, 1961), pp. 153–155.

> Assuming that Wilberforce correctly represented Norris's views, this statement is an accurate inference from the quotation presented here. It would be preferable, however, to begin the sentence, "According to Wilberforce . . . ". The quote is correctly footnoted.

William Wilberforce did not object to the slave trade per se, but he believed that the conditions of transit were a problem, "the most wretched part of the whole subject."

> This statement is an inference from silence, based upon a quotation taken out of context. The quotation gives no direct evidence of Wilberforce's opinion of slavery. His opposition can be inferred from the quotation, in that the phrase "most wretched" implies that other features of slavery were "wretched," too.

Wilberforce reported that it was the constant practice of slave-traders to set sail "in the night lest they [the slaves] should be sensible of their departure."

> The documentation for this assertion is inadequate because, while the evidence is correct and the conclusions drawn are correct, parts of the quotation are presented without quotation marks. In addition, no citation is given for the quoted text.

In contrast, Mr. Norris, a Liverpool trader, said that the slaves are not fettered at the wrist, even "if they are turbulent."

> The quotation is correct but the facts (assuming that Wilberforce is accurately quoting Norris) are wrong: slaves *were* fettered at the wrist if they were "turbulent." Also, no citation is given for the quoted text.

Wilberforce accused Norris of lying to the Privy Council about the conditions of slave transportation because he did not want any government interference in the trade.

> Wilberforce *may* have thought Norris was lying, but his speech here does not say so. What Wilberforce says is that "interest" (in Norris's case, his trading business) nearly blinds people to the truth. See lines 9–13.

The slave traders tried to take good care of their human cargo: they provided them with good food, exercise, and entertainment on board the ships that carried them to the West Indies.

> The traders *may* have tried to take good care of their cargo. (They were, after all, an investment.) But the text offers two views of their treatment,

and the accuracy of either as fact is disputable because of the existence of the other and the lack of citation to corroborating sources.

Wilberforce said that the slave quarters on board ship were scented with "frankincense and lime-juice" and their food included "the best sauces of African cookery."

It is technically correct to say that Wilberforce "said" these things, but a falsehood is the result, because of course Wilberforce is representing, in these quotations, the words and opinions of another person. Rather than "said," "reported" would be a better term to use here. In addition, the statement should be prefaced and followed by other statements making it clear that Wilberforce is not making these assertions himself, but is reporting the assertions of others.

In order to keep the slaves healthy, the slave traders hired men who whipped the slaves to force them to exercise.[2]

[2]Ibid.

This statement incorrectly implies that slaves were always whipped to force them to exercise. Wilberforce did not say that; he said the threat of the whip and "sometimes . . . the actual use of it" made the slaves dance. See lines 42–43. The statement does, however, correctly cite its source.

The slaves did not want to exercise and were forced to do so by their owners.[3]

[3]Ibid.

There is no evidence here about what the slaves wanted or didn't want. For instance, they may have wanted to exercise, but not while shackled to another person. The writer here assumes the wishes of the slaves from the threats of whipping described by Wilberforce.

Wilberforce's racism can be seen in the way he likened African slaves to cattle, "sound in wind and limb."

Here the context is ignored. Wilberforce likens the slave-buyer's purchase of healthy slaves to the farmer's purchase of cattle. Wilberforce's own opinion is not discernible from this quotation. In addition, no citation is given for the quoted text.

Healthy Africans, after spending some time on slave ships in transit to the New World, came down with diseases that often killed them.[4]

[4]Ibid.

The quotation presented here, with the reference to the purchase of healthy slaves and their sickness on board ships, allows this inference. However, the implication is that the conditions of the ships *caused* the diseases, and that cannot be established from this quotation. Some diseases are not visible in their early stages, and they could be said to have been transmitted because of the slave trade and the close quarters that such a trade enforced. Again, it would be better to begin the sentence, "According to Wilberforce . . . ".

The death rate for slaves on the trip from Africa to the New World was 50%.

This is a correct description of *Wilberforce's* statement, but it is not necessarily true. In any case, the assertion should be footnoted.

Recent historians have found that the mortality of Africans being transported to the West Indies varied from 10% to 22%.[5]

[5]Philip D. Curtin, *The Atlantic Slave Trade: A Census* (Madison: University of Wisconsin Press, 1969), p. 279.

This sentence is an appropriately footnoted use of a secondary source.

Wilberforce inflated the mortality statistics in order to strengthen his case against the slave trade.

There is no evidence in the text to suggest that Wilberforce knowingly inflated the mortality figures. Nor is there evidence, in the text provided, that he did not.

Recently historians have argued that the mortality rates on slave ships were worse for white seamen than they were for the slaves being transported.[6]

[6]Ibid., pp. 279, 282–283.

This sentence is an appropriately footnoted use of a secondary source.

Had Wilberforce known this it would have destroyed his arguments against the slave trade, because he thought "Death . . . is a sure ground of evidence."

No. The text shows that Wilberforce thought that the death rates were the strongest evidence he could offer for the real problem of the transport: "misery." See lines 49–53. But this misery took many forms: bad food, little water, crowding, stench, the lash, cruel treatment, grief for home. Again, however, the assertion and quoted text should be footnoted.

18. Oral Historians Meet the Media

While the case presented in chapter 18 raises issues of editorial control over the publication of historical research, the process of interviewing for historical documentation also raises ethical issues.

Some background information may be useful. Oral history interviews might be conducted for either of two purposes. The first, archival oral history, involves interviews conducted for preservation purposes. Transcripts of the interviews are housed in a library or archive for future use by scholars. In contrast, interviewing for historical documentation refers to interviews conducted by scholars to obtain answers to questions of particular interest to them. Subjects of archival oral histories are given the opportunity to review the transcripts of their interviews and change or withdraw statements they wish to correct or withhold. Wherever possible, interviewers inform subjects of contradictions or discrepancies either within their own interviews or with other evidence. The purpose of this review and correction process is to produce a record that is the result of some amount of reflection, thereby giving subjects the opportunity to be as accurate as possible and also to protect them from unintentionally divulging sensitive information. It also gives subjects some control over the interview as, for example, when the interview spills over its previously agreed-upon boundaries.

Many of the issues associated with oral historical research are common to all research involving human subjects. (See, for example, chapters 10 and 11.) These issues, presented here in the context of conducting oral history interviews, can be framed as follows:

1. What information must the researcher provide to the subject so that he or she can make an informed, voluntary choice to grant the interview?

2. Should researchers be required to disclose all of their purposes and motivations in their request to interview a subject? Why or why not?

3. Should researchers ask questions that go beyond the stated purposes of their interviews with subjects? Is it ethical, for example, for researchers conducting interviews of subjects who have given permission to be interviewed about their public lives to ask questions about their personal lives as well?

4. Should the scope of the questions that researchers ask subjects depend on whether the subject is a public figure? Are there limits to the types of questions that should be asked of public figures?

5. What similarities and differences do you see between oral history and journalism? Are these similarities and differences ethically relevant?

6. Of what relevance, if any, is the interviewer's training in conducting oral histories to the interviewer's obligation to follow accepted professional standards?

Another set of questions that might be raised involves the reliability of the information gained through oral interviews and the uses to which such information can justifiably be put.

1. What is the relative weight that should be given to the various available documentary sources, interviews as well as other archival and secondary resources?

2. Should oral historians be obligated to bring factual disparities between a subject's recollection and prior testimony by the subject or from other sources to the subject's attention during the interview? Why or why not?

3. Should oral historians allow their subjects to edit the transcripts of their oral interviews to "correct" for such disparities before publication? Why or why not? Are there other kinds of "corrections" that subjects should be allowed to make? What special problems might such a policy present with, for instance, elderly subjects?

You may also want to discuss with your students the problems associated with researcher-subject relationships. It is not uncommon for the relationship between a researcher and subject to develop aspects extending beyond a formal professional relationship. Often, researchers deliberately attempt to build a relationship of trust and friendship with their subjects in order to put them at ease and make them comfortable sharing personal information they might otherwise be reluctant to disclose. Sometimes a relationship of trust or dependency may arise naturally during the course of the interviews. At some point, however, the historian-subject relationship may be stretched beyond the limits to which it should appropriately extend.

You may wish to raise the following questions for discussion:

1. What obligations, if any, should oral historians have to protect the safety and well-being of their subjects? What obligations, if any, do they have to refrain from personal involvement with their subjects?

2. Does your answer to Question 1 depend on the circumstances of a particular case? If so, what considerations are relevant to the determination of whether, and to what extent, an obligation exists (e.g. the subject's age, mental capacity, the existence of close family members or friends, the number and duration of the interview sessions)?

3. What obligations, if any, should researchers have to clarify the nature (including the duration and depth) of their intended involvement with subjects at the inception of the interviews?

4. How should researchers handle the dependency that may result from the relationships that develop with subjects as a result of the interview situation?

5. What obligations, if any, should researchers have to safeguard the privacy of subjects who may become public figures as a result of being interviewed?

The American Historical Association, in consultation with the Oral History Association, the Organization of American Historians, and the Society of American Archivists, has developed guidelines for interviewing for historical documentation. They are reprinted here for your consideration.[9]

Statement on Interviewing for Historical Documentation

Interviewing has become commonplace in historical research focusing on the twentieth century, but unfortunately it is often done and used without proper attention to professional obligations. When they conduct interviews, individual historians too often fail to adhere to the standards now well-established in more formal oral history programs and projects. Historians should recognize that in interviewing they are creating historical documents, and that entails special responsibilities to ensure future access for both verification and research by others. The AHA's Statement on Standards of Professional Conduct establishes basic obligations for historians who engage in interviewing:

> Historians should carefully document their findings and thereafter be prepared to make available to others their sources, evidence, and data, including the documentation they develop through interviews. . . .
>
> Since historians must have access to sources—archival and other—in order to produce reliable history, they have a professional obligation to preserve sources and advocate free, open, equal, and non-discriminatory access to them, and to avoid actions which might prejudice future access. Historians recognize the appropriateness of some national security and corporate and personal privacy claims but must protect research collections and other historic resources and make those under their control available to other scholars as soon as possible.
>
> Certain kinds of research and conditions attached to employment or to use of records impose obligations to maintain confidentiality, and oral historians often must make promises to interviewees as conditions for interviews. Scholars should honor any pledges made. At the same time historians should seek definitions of confidentiality before work begins, press for redefinitions when experience demonstrates the unsatisfactory character of established regulations, and advise their readers of the conditions and rules that govern their work. They also have the obligation to decline to make their services available when policies are unnecessarily restrictive.

Recognizing the need for more specific guidelines, the Association's Professional Division consulted with representatives of the Oral History Association, the Organization of American Historians, and the Society of American

Archivists. The following guidelines resulted from that discussion and are drawn from statements adopted by the Oral History Association and the Society for History in the Federal Government:

1. Interviews should be recorded on tape but only after the person to be interviewed has been informed of the mutual rights and responsibilities involved in oral history, such as editing, confidentiality, disposition, and dissemination of all forms of the record. Interviewers should obtain legal releases and document any agreements with interviewees.
2. The interviewer should strive to prompt informative dialogue through challenging and perceptive inquiry, should be grounded in the background and experiences of the person being interviewed, and, if possible, should review the sources relating to the interviewee before conducting the interview.
3. To the extent practicable, interviewers should extend the inquiry beyond their immediate needs to make each interview as complete as possible for the benefit of others.
4. The interviewer should guard against possible social injury to or exploitation of interviewees and should conduct interviews with respect for human dignity.
5. Interviewers should be responsible for proper citation of oral history sources in creative works, including permanent location.
6. Interviewers should arrange to deposit their interviews in an archival repository that is capable of both preserving the interviews and making them available for general research. Additionally, the interviewer should work with the repository in determining the necessary legal arrangements.
7. As teachers, historians are obligated to inform students of their responsibilities in regard to interviewing and to encourage adherence to the guidelines set forth here.

See also the *Oral History Evaluation Guides*, published by the Oral History Association, and John Neuenschwander's *Oral History and the Law* (Oral History Association Pamphlet No. 1, 1985), which provides sample release forms.

19. Limiting Access to Scholarly Materials: The Case of the Dead Sea Scrolls

This set of cases is designed to explore three basic issues:

1. What are the conditions under which one can restrict or accept restrictions on research activities for the betterment of the profession?

2. How much time can ethically be taken to complete a project that is restricted? Discussion should include the duty of the historian to get the material out in a reasonable period of time. Determining the "reasonableness" of that time may prove difficult. In the context of the Dead Sea Scrolls, some scholars have argued that the importance of the documents to our understanding of two of the world's major religions requires meticulous work that should not be shared until the scholars working on the documents are as sure as they can be that they've gotten it right.

3. Were the actions of the scholars who produced the reconstruction from the concordance ethically justified?

There are a number of unresolved factual issues and problems presented by this case, of which you may want to make students aware:

1. What criteria were used by Jordanian authorities in appointing the initial editorial board, which included no Jewish or Israeli scholars?

2. When they acquired jurisdiction over the scrolls, why did the IAA continue with the same editorial board and board policy that the Jordanians had instituted?

3. What were the circumstances under which Ms. Bechtel gained permission to have photographs of the documents made? Did she agree to restrict others' access to these photographs? What was the nature of her dispute with the Claremont Institute and her subsequent donation of the scrolls to the Huntington Library? Of what effect is any agreement Ms. Bechtel had with the IAA, regarding restrictions, on her agreement with the Huntington?

You might think about juxtaposing the facts of the Dead Sea Scrolls case with those of other examples of restricted access to research materials, such as:

1. classified research;
2. sensitive biographical materials;
3. interviews in which the interviewee wishes to remain unknown;
4. archives to which access is limited; and
5. private letters.

At about the same time as the events described in this case, another scholar, Robert Eisenman, was active in the move to secure scholarly access to unpublished Dead Sea Scroll materials. He subsequently published two works, the introductions to which raise interesting questions regarding restricted access to scholarly materials, particularly the volume co-edited with Michael Wise (see Eisenman and Wise 1991: 1–16 and Eisenman and Robinson 1991: ix–xi and xiii-xiv).

Part A

Question 1. The IAA attempted to legitimate continuing the policy of exclusive access to the scrolls based on its interests in authorizing an "official version" in order to prevent shoddy scholarship. Is this a persuasive rationale? Why or why not? Note that there are two notions of an "official version": one denotes an authorized version based on excellent scholarship; the other refers to a version in line with a particular set of beliefs. Which, if either, of these meanings of "official" would be a legitimate reason for restricting access to the scrolls?

Question 2.

1. Do certain materials provide information that is so valuable that exclusive access cannot be justified? Do such circumstances make the "quality control" argument more or less persuasive?

2. From a different perspective, are certain materials so precious and valuable that access to the originals should be severely restricted in order to preserve the quality of the materials? Distinguish between access to the original scrolls or scroll fragments and access, even to facsimiles, that will result in publication of "unauthorized" texts.

Question 3. What difference, if any, should ownership as opposed to possession make in terms of the legitimacy of those controlling research materials to restrict access to them? What difference would it make if some of the scrolls in the possession of the Israelis (and before them, the Jordanians), had not been purchased from their Bedouin discoverers, but had instead been confiscated from them or from middlemen? Is the nature of the documents relevant to your assessment (e.g., the difference between private diaries and religious or legal manuscripts)? Why or why not?

Question 4. What difference, if any, should private or public ownership of research materials make in terms of the legitimacy of restricting access? Why?

Part C

Question 2. A number of follow-up questions might be asked regarding the Huntington's ethical obligation, including:

> 1. Private libraries, such as the Huntington, often house materials access to which is restricted by the terms of the donation. What implications, if any, does the Huntington's decision to open up access to the scrolls have, both for the Huntington's own access policies, and for those of other, similar institutions? What effect might the Huntington's decision have on future donations to its collections?
>
> 2. To follow up on the question of prior arrangements between the IAA and Ms. Bechtel, what if the IAA had only agreed to allow the photographs to be made on the condition that they not be distributed beyond

the designated places of safekeeping? What if Ms. Bechtel had made the second copy of the photographs without the prior permission of the IAA?

Part D

Question 2. What effect should privacy and private property interests have in limiting rights to access research materials? Where should the line between public access and protection of privacy interests be drawn? Does it depend on the nature of the materials at issue, that is, their value to the general public as opposed to a small group of scholars?

20. Faculty–Graduate Student Relations

Case 1

In leading the discussion of this case, you would want to draw out whether Linda has misunderstood what is going on (is Professor Crane really having the students write his book chapters for him); what kind of credit she would get for her work; what of her work, if any, he intends to use; and whether or not she has any choice (i.e., can she choose to do a topic of her own choice or must she do one of his "chapters").

21. Intellectual Property

Case 7

The problems most often associated with masked reviews are accountability on the part of the referee and the ability of the author to trace appropriations of his or her work. When the author is known to the referee, however, the author's reputation and importance in the field may unduly influence the referee's judgment (either positively or negatively). Revealing the identity of the referee to the author may have a chilling effect on the referee's willingness to give a candid review. Many publishers allow the referee to decide whether or not his or her identity and comments (one or the other or both) may be given to the author.

Notes

1. 46 N.Y.2d 401, 413 N.Y.S.2d 895, 836 N.E.2d 807 (1978).
2. See, e.g., *Sherlock v. Stillwater Clinic*, 260 N.W.2d 169 (Minn. 1977); *Troppi v. Scarf*, 31 Mich. App. 2450, 187 N.W.2d 511 (1971).
3. ______ U.S. ______, 111 S.Ct. 1196, 113 L.Ed. 158 (1991).
4. 45 *Code of Federal Regulations* 46, Subpart D.
5. U.S. Department of Agriculture. "Rules and Regulations." *Code of Federal Regu-*

lations Title 9, Part 2, Subpart C, Section 2.31. "Institutional Animal Care and Use Committee (IACUC)."

6. Stanley Milgram, "Some Conditions of Obedience and Disobedience to Authority," *Human Relations* 18: 57–76 (1965); Idem, "Behavioral Study of Obedience," *Journal of Abnormal and Social Psychology* 6(4): 371–378 (1963). Criticisms of Milgram's work and some of Milgram's responses to his critics may be found in Diana Baumrind, "Research Using Intentional Deception," *American Psychologist* 40(2): 165–174 (February 1985); Stanley Milgram, *Obedience to Authority: An Experimental View*, pp. 193–202 (New York: Harper & Row, 1974); Idem, "Issues in the Study of Obedience: A Reply to Baumrind," *American Psychologist* 19: 848–852 (1964); Arthur G. Miller, *The Obedience Experiments: A Case Study of Controversies in Social Science* (New York: Praeger, 1986).

7. Federal regulations define "minimal risk" to mean "that the probability and magnitude of harm or discomfort anticipated in the research are not greater in and of themselves than those ordinarily encountered in daily life or during the performance of routine physical or psychological examination or tests" (45 *Code of Federal Regulations* 46.102[i]).

8. A good discussion of this question is provided in U.S. Department of Health and Human Services. (1993), pp. 3–44 to 3–46 and 5–65.

9. American Historical Association. "Statement on Interviewing for Historical Documentation." In *Statement on Standards of Professional Conduct* (Washington, D.C.: American Historical Association, 1992), pp. 25–27. Reprinted with permission. Copies of this and other AHA statements are available from the American Historical Association, 400 A St., S.E., Washington, DC 20003-2422.

References

Angier, Natalie. "Team Pinpoints Genetic Cause of Huntington's." *New York Times* (March 24, 1993), p. 1.

Clingerman, Karen. "Ethical and Moral Issues Relating to Animals 1979–1988." *Quick Bibliography Series*. Beltsville, MD: U.S. Department of Agriculture, 1988).

Donnelly, Strachan, and Nolan, Kathleen. "Animals, Science, and Ethics." *Hastings Center Report* 20(3): 1–32 (Special Supplement May/June 1990).

Eisenman, Robert H., and Wise, Michael, eds. *The Dead Sea Scrolls Uncovered: The First Complete Translation and Interpretation of 50 Key Documents Withheld for Over 35 Years*. London and New York: Penguin Books, 1991.

——, and Robinson, James M., eds. *A Facsimile Edition of the Dead Sea Scrolls*. 2 vols. Wash. DC: Biblical Archaeology Society, 1991.

Freedman, Benjamin. "Equipoise and the Ethics of Clinical Research." *New England Journal of Medicine* 310(3): 141–145 (July 16, 1987).

Herzog, Harold. "Discussing Animal Rights and Animal Research in the Classroom." *Teaching of Psychology* 17: 90–94 (1990).

Huntington's Disease Collaborative Research Group. "A Novel Gene Containing a Trinucleotide Repeat That Is Expanded and Unstable on Huntington's Disease Chromosomes." *Cell* 72: 971–983 (1993).

McCormick, Richard. *How Brave a New World? Dilemmas in Bioethics*. Garden City, NY: Doubleday, 1981.

Pihl, R. O.; Smith, Mark; and Farrell, Brian. "Alcohol and Aggression in Men: A Comparison of Brewed and Distilled Beverages." *Journal of Studies on Alcohol* 45: 278–282 (1984).

Ramsey, Paul. *The Patient as Person*. New Haven: Yale University Press, 1970.

——. "The Enforcement of Morals: Nontherapeutic Research on Children." *Hastings Center Report* 6: 21–39 (August 1976).

U.S. Department of Health and Human Services. "Protection of Human Subjects." *Code of Federal Regulations.* Title 45, Part 46 (1990).

U.S. Department of Health and Human Services. Public Health Service. National Institute on Alcohol Abuse and Alcoholism. National Advisory Council on Alcohol Abuse and Alcoholism. "Recommended Council Guidelines on Ethyl Alcohol Administration in Human Experimentation." Revised June 1989.

U.S. Department of Health and Human Services. Public Health Service. National Institutes of Health. Office for Protection from Research Risks. *Protecting Human Research Subjects: Institutional Review Board Guidebook.* Washington, DC: Government Printing Office, 1993.

Veatch, Robert. *The Patient as Partner: A Theory of Human Experimentation Ethics.* Bloomington: Indiana University Press, 1987.

Weaver, Robert F., and Hedrick, Philip W. *Genetics.* Duquesne, IA: William C. Brown, 1989.

Annotated Bibliography

Compiled by Lucinda Peach

Contents

General Issues

Biology

1. Chapter 3, The Professional Scientist; Chapter 4, Scientific Misconduct: What Is It and How Is It Investigated?; Chapter 5, Authorship and the Use of Scientific Data; Chapter 6, Data Alteration in Scientific Research
2. Chapter 7, The Ethics of Genetic Screening and Testing
3. Chapter 8, Ethics and Eugenics

Psychology

1. Chapter 9, Ethical Issues in Animal Experimentation
2. Chapter 10, Research Involving Human Subjects; Chapter 11, Research Involving Human Subjects: The Administration of Alcohol; Chapter 12, The Ethics of Deception in Research
3. Chapter 13, Misconduct in Science

History

1. Chapter 16, The Historian's Code of Ethics; Chapter 17, The Use and Interpretation of Historical Documents; Chapter 21, Intellectual Property
2. Chapter 18, Oral Historians Meet the Media
3. Chapter 19, Limiting Access to Scholarly Materials: The Case of the Dead Sea Scrolls

General Issues

Council of Biology Editors. Editorial Policy Committee, eds. *Ethics and Policy in Scientific Publication.* Bethesda, MD: Council of Biology Editors, Inc., 1990. (A collection of scenarios dealing with various ethical issues faced by publishers of scientific journals. Includes discussion of the issues presented as reported by members of the Council of Biology Editors in response to a survey of members and analysis of the responses by the volume's editors.)

Friedman, Paul J., ed. "Integrity in Biomedical Research." *Academic Medicine* 68(9) (Special Supplement September 1993). (An excellent collection of articles by noted authors and activists in the field is presented in this special supplement devoted to the topic of scientific integrity. A broad range of topics is covered.)

Grinnell, Frederick. *The Scientific Attitude.* Boulder, CO: Westview Press, 1987.

Jonsen, Albert. "Of Balloons and Bicycles—or—the Relationship between Ethical Theory and Practical Judgment." *Hastings Center Report* 21: 14–16 (September/October 1991). (Analogizes the relationship between moral theory and practical judgments in particular cases to the metaphor of a hot-air balloon and a bicycle, suggesting that theory usually is not necessary but can provide direction in unfamiliar territory.)

Jonsen, Albert R., and Toulmin, Stephen. *The Abuse of Casuistry: A History of Moral Reasoning*. Berkeley: University of California Press, 1988. (Chapter 16, "The Revival of Casuistry," provides general background about the casuistical method of moral reasoning, including illustrations of how it functions when applied to particular case examples.)

Macrina, Francis L., and Munro, Cindy L. "Graduate Teaching in Principles of Scientific Integrity." *Academic Medicine* 68(12): 879–886 (December 1993). (Describes the development of a course on the responsible conduct of research at Virginia Commonwealth University. An annotated list of resources on scientific integrity follows at pp. 885–886.)

May, William. "Doing Ethics: The Bearing of Ethical Theories on Fieldwork." *Social Problems* 27: 358–370 (1980). (Discusses the relevance of teleological, utilitarian, deontological, critical philosophical, and conventional ethical theories to fieldwork.)

National Academy of Sciences. *On Being a Scientist*. Washington, DC: National Academy Press, 1989. (Excellent introductory primer for beginning scientists. Describes some basic aspects of scientific research and the ethical problems that arise in research, including the treatment of data, self-deception, peer review, fraud, allocation of credit and responsibility in collaborative research, plagiarism, and upholding the integrity of science.)

National Academy of Sciences. National Academy of Engineering. Institute of Medicine. Committee on Science, Engineering, and Public Policy. Panel on Scientific Responsibility and the Conduct of Research. *Responsible Science: Ensuring the Integrity of the Research Process*. 2 vols. Washington, DC: National Academy Press, 1992. (The report of COSEPUP's Panel on Scientific Responsibility and the Conduct of Research, which was convened in 1989 to study the factors that affect the responsible conduct of research and to make recommendations for maintaining and improving the integrity of the research process. The report also addresses the processes for handling allegations of misconduct. Volume I comprises the findings and recommendations of the study panel. Volume II includes the panel's working papers and various institutional policies that the panel found useful to its work.)

Nova. "Do Scientists Cheat?" Northbrook, IL: Coronet Film and Video, 1988. (Videotape.) (Examines why scientific fraud is so hard to detect and details the numerous factors that influence fraud, including an increasing competition for grant money and tenured positions. Discusses why scientists may be less than honest, analyzes how our scientific system deals with quality control, and considers the adequacy of the scientific community's response when a researcher is involved in fraud.)

Reiser, Stanley Joel, and Heitman, Elizabeth. "Creating a Course on Ethics in the Biological Sciences." *Academic Medicine* 68(12): 876–879 (December 1993). (Describes the development of a course on the responsible conduct of research. An annotated list of resources on scientific integrity follows at pp. 885–886.)

Sachs, G. A., and Siegler, Mark. "Teaching Scientific Integrity and the Responsible Conduct of Research." *Academic Medicine* 68(12): 871–875 (December 1993). (Describes the development of a course at the University of Chicago's medical school on the responsible conduct of research. An annotated list of resources on scientific integrity follows at pp. 885–886.)

Sigma Xi, the Scientific Research Society. *Honor in Science*. New Haven, CT: Sigma Xi, the Scientific Research Society, 1986. (An excellent introductory pamphlet of practical advice for beginning researchers in science. Discusses ethical considerations relevant to research, including honesty, manipulation of data, collaborative research, customary practices in the scientific community as a standard for assessing misconduct, and responsibilities for reporting misconduct and how to seek assistance in doing so.)

Teich, Albert, and Frankel, Mark. 1992. *Good Science and Responsible Scientists: Meeting the Challenge of Fraud and Misconduct in Science*. Washington, DC: American Association for the Advancement of Science, 1992. (A useful overview of the problems of research fraud and misconduct in science. Explains the ambiguity in definitions of fraud and misconduct, summarizes the notorious cases of research misconduct which provided the impetus for governmental regulation, the structure of federal regulation and oversight, the role of journals, scientific societies and research institutes, and the problems encountered in whistleblowing.)

U.S. Department of Health and Human Services. Public Health Service. "Responsibilities of Awardee and Applicant Institutions for Dealing with and Reporting Possible Misconduct in Science." *Federal Register* 54(151): 32446–32451 (August 8, 1989).

Biology

1. Chapter 3, The Professional Scientist; Chapter 4, Scientific Misconduct: What Is It and How Is It Investigated?; Chapter 5, Authorship and the Use of Scientific Data; Chapter 6, Data Alteration in Scientific Research

SHORTER REFERENCES OF PRIMARY INTEREST

General

Barber, Bernard. "Trust in Science." *Minerva* 25: 123–134 (Spring/Summer 1987). (Discusses the role of trust—in both the senses of trustfulness and trustworthiness—as foundational to the enterprise of science and scientific research, the fiduciary duties of scientists, fraud, the public's trust in science, and deception of human subjects.)

Branscomb, Lewis. "Integrity in Science." *American Scientist* 73: 421–423 (1985). (A personal perspective on the costs that researchers' lack of integrity and honesty have on the scientific community as a whole.)

Chubin, Daryl. "Research Malpractice." *Bioscience* 35: 80–89 (1985). (A good general overview of the area of research malpractice, divided into the areas of production of research, reporting, dissemination, and evaluation. Chubin identifies "normal" and "deviant" behaviors in each phase, and offers suggestions for reducing research malpractice.)

Roman, Mark. "When Good Scientists Turn Bad." *Discover* 9(4): 51–58 (April 1988). (Describes recent instances of scientific misconduct, including the cases involving Stephen Breuning and Charles Glueg.)

Whistleblowing

Bok, Sissela. "Whistleblowing and Professional Responsibilities." In *Ethics Teaching in Higher Education*, edited by Daniel Callahan and Sissela Bok, pp. 277–295. New York: Plenum Press, 1980. (Discusses the various responsibilities that confront would-be whistleblowers, using examples drawn from government, business, and engineering, and proposes a casuistical method of resolving the conflicts faced by whistleblowers. Also recommends analyzing problems of whistleblowing from the perspective of institutions and professions.)

Sprague, Robert. "A Case of Whistleblowing in Research." *Perspectives on the Professions* 8: 4–5 (1989). (A brief account of the author's experiences in "blowing the whistle" on the research misconduct of then-colleague Stephen Breuning.)

Conflicts of Interest (Academic Responsibilities versus Commercial Opportunities)

Bok, Derek. "Balancing Responsibility and Innovation." *Change* 14: 16–25 (1982). (An excellent discussion of the issues faced by a university attempting to negotiate technology transfer and to control the growth and direction of its relationship with industry.)

Kenny, Martin. *Biotechnology: The University-Industrial Complex*. New Haven, CT: Yale University Press, 1986. (Pages 113–125 provide a general background discussion of conflicts of interest that arise in connection with university researchers working for commercial enterprises, including impact on graduate students and the free flow of information.)

Authorship

Burman, Kenneth. " 'Hanging from the Masthead': Reflections on Authorship." *Annals of Internal Medicine* 97: 602–605 (1982). (Criticizes the general practice of multiple authorship of research articles, which does not accurately represent actual involvement in the underlying research or writing.)

Campbell, Donald. "Ethical Issues in the Research Publication Process." In *Ethical Dilemmas for Academic Professionals*, edited by S. L. Payne and B. H. Charnov, pp. 69–85. Springfield, IL: Charles C. Thomas, 1987. (A general overview of a number of issues involved in publishing research, including plagiarism, multiple authorship, public review process, faculty obligations to students, and distortion of findings.)

Croll, Roger. "The Noncontributing Author: An Issue of Credit and Responsibility." *Perspectives in Biology and Medicine* 27: 401–407 (1984). (Exposes inaccuracies in the assumption that all authors listed on a research article have contributed to the underlying research, discusses problems with current practices, and proposes changes that would require authors to have made significant contributions to the research and make them responsible for all aspects of the article.)

Editorial. "Why Scientific Fact Is Sometimes Fiction." *Economist* 302: 97–98 (1987). (A brief but informative overview of the recent spate of misconduct cases in published research, citing inadequate referee review, busy laboratory supervisors, large research teams, and multiple authorship as some of the problems which have led to this result.)

Faculty Obligations

Campbell, Donald. "Ethical Issues in the Research Publication Process." In *Ethical Dilemmas for Academic Professionals*, edited by S. L. Payne and B. H. Charnov, pp.

69–85. Springfield, IL: Charles C. Thomas, 1987. (A general overview of a number of issues involved in publishing research, including plagiarism, multiple authorship, public review process, faculty obligations to students, and distortion of findings.)

Brown, Robert, and Krager, LuAnn. "Ethical Issues in Graduate Education: Faculty and Student Responsibilities." *Journal of Higher Education*. 56: 403–418 (1985). (Provides a framework for assessing the ethical dimensions in both faculty and student responsibilities in academia. Applies the ethical principles of autonomy, nonmaleficence, beneficence, justice, and fidelity to the faculty member's roles as advisor, instructor, curriculum planner, researcher, and mentor, and the student's reciprocal roles.)

Intellectual Property

Heathington, K. W., et al. "Commercializing Intellectual Properties at Major Research Universities: Income Distribution." *Journal of Social Research Administration* 17: 27–39 (1986). (Surveys university policies regarding income sharing from profits derived from patents and copyrights, increasing collaboration between university researchers and commercial enterprises, and changes in federal policy and regulation which have enhanced the opportunities and desirability of such collaboration.)

Linnell, Robert H. "Intellectual Property: Developing an Equitable Policy." *Business Officer* (March): 19–24 (1983). (A carefully considered investigation of the major ethical, legal, and policy considerations and institutional policies and practices surrounding profit sharing from commercial applications of intellectual property developed by university faculty.)

U.S. Department of Health and Human Services. Public Health Service. Office of the Assistant Secretary of Health, Office of Health Planning and Evaluation, and Office of Scientific Integrity Review. "Data Management in Biomedical Research: Report of a Workshop." (April 1990). (Provides useful discussion on issues of data ownership, sharing, and retention.)

Zatz, Joel. "Intellectual Property: An Academician's Perspective." *American Journal of Pharmaceutical Education* 53: 346–350 (1989). (Excellent general background to the issues surrounding patents in the academic context, including faculty research support, publication, students, record-keeping, and confidentiality.)

LONGER REFERENCES AND REFERENCES OF SECONDARY INTEREST

Bok, Derek. *Beyond the Ivory Tower: Social Responsibilities of the Modern University*. Cambridge, MA: Harvard University Press, 1982. (Chapter 6, "Academic Science and the Quest for Technological Innovation," covers the conflicts of interest that arise within the university between academic integrity and commercialization of research.)

Bok, Sissela. *Lying*. New York: Vintage Books, 1989. (A general treatment of the topic of lying, including whether the "whole truth" is attainable; the relationship between truthfulness, deceit, and trust; the consequences of lying; excuses; justifications; lying to protect others or for the "public good"; deceptive social science research; and "paternalistic lies.")

———. *Secrets*. New York: Vintage Books, 1989. (A general treatment of the topic of secrecy, including different approaches, secrecy's relation to moral choice, secrecy and self-deception, power and accountability, intrusive social science research, and the limits of confidentiality. Of particular relevance are chapters 11, "Secrecy and Competition in Science," and 14, "Whistleblowing and Leaking.")

Branscomb, Anne. "Who Owns Creativity? Property Rights in the Information Age." *Technology Review* 91: 39–45 (1988). (Discusses ethical dilemmas in the area of intellectual property that have resulted from new electronic technologies such as computer programs. Includes a general background of intellectual property laws.)

Goldman, Alan. "Ethical Issues in Proprietary Restrictions on Research Results." *Science, Technology, and Human Values* 12: 22–30 (1987). (A sophisticated analysis of the value issues surrounding patents in the academic context. Provides an analysis of the arguments for and against patent ownership by faculty researchers.)

Hanson, Shirley. "Collaborative Research and Authorship Credit: Beginning Guidelines." *Nursing Research* 37: 49–52 (1988). (Synthesizes studies on collaborative research and multiple authorship to derive a number of practical principles for how to proceed with such research.)

LaFollette, Marcel C. "Beyond Plagiarism: Ethical Misconduct in Scientific and Technical Publishing." *Book Research Quarterly* 4: 65–73 (1988–89). (Analyzes the growing problem of scientific misconduct from a historical perspective, tracing changes in the nature of the scientific community starting with World War II and the growing involvement of the government in research, including regulation.)

———. *Stealing into Print: Fraud, Plagiarism, and Misconduct in Scientific Publishing.* Los Angeles: University of California Press, 1992.

Nelkin, Dorothy. *Science as Intellectual Property: Who Controls Research?* New York: Macmillan, 1983. (Discusses the growing commercialization of research, conflicts of interest between faculty members' commercial and academic responsibilities, increased regulation, whistleblowing, and various disputes involving control of and access to scientific information.)

U.S. House of Representatives. Subcommittee on Investigations and Oversight. Committee on Science, Space, and Technology. *Maintaining the Integrity of Scientific Research.* Washington, D.C.: U.S. Government Printing Office, 1990. (Reports on the investigation of ten cases of fraud and misconduct in scientific research, including the much-publicized cases involving Breuning, Darsee, and Slutsky, with additional findings, conclusions, and recommendations for policing scientific fraud and misconduct.)

Woolf, Patricia. "Deception in Scientific Research." *Jurimetrics Journal* 29: 67–95 (Fall 1988). (Provides an interesting discussion of misconduct in research, including a summary of a number of fairly recent cases, how they were dealt with, and how the wrongdoers were disciplined.)

2. *Chapter 7, The Ethics of Genetic Screening and Testing*

SHORTER REFERENCES OF PRIMARY INTEREST

American Medical Association. Council on Ethical and Judicial Affairs. "Use of Genetic Testing by Employers." *Journal of the American Medical Association* 266: 1827–1830 (1991). (Considers ethical problems with workplace testing of employees, either to protect worker safety or avoid increased health care costs by excluding those who have increased risks of disease from exposure to workplace hazards or to protect public safety.)

Angier, Natalie. "Team Pinpoints Genetic Cause of Huntington's." *New York Times* (March 24, 1993), p. 1.

Annas, George. "Mandatory PKU Screening: The Other Side of the Looking Glass." *American Journal of Public Health* 72: 1401–1403 (1982). (The arguments propounded in this article are contrary to the policy of most states, which have enacted mandatory PKU screening tests, and to the position argued in the Faden et al. article cited below. Annas concludes that PKU screening programs for newborns should be voluntary and require parental consent.)

Baum, Rudy. "Genetic Screening: Medical Promise among Legal and Ethical Questions." *Chemical and Engineering News* 67: 10–16 (January 19, 1986). (Excellent overview of genetic screening, how RFLP analysis works, and the relevant ethical issues, particularly mandatory screening in employment and prenatal contexts.)

DeGrazia, David. "The Ethical Justification for Minimal Paternalism in the Use of the Predictive Test for Huntington's Disease." *Journal of Clinical Ethics* 2(4): 219–228 (Winter 1991).

Faden, Ruth, et al. "Parental Rights, Child Welfare, and Public Health: The Case of PKU Screening." *American Journal of Public Health* 72: 1396–1400 (1982). (Contrary to the Annas article cited above, Faden et al. conclude that moral considerations weigh in favor both of making PKU screening of newborns mandatory and of not requiring parental consent.)

Gustafson, James. "Genetic Counseling and the Uses of Genetic Knowledge—An Ethical Overview." In *Ethical Issues in Human Genetics*, edited by Bruce Hilton et al., pp. 101–112. New York: Plenum Press, 1973. (A philosophical treatment of the ethical issues involved in genetic counseling relating to genetically impaired fetuses. Poses the central issue as whether the individual or the community should have primacy in ethical considerations. Discusses the tension between individual rights-based approaches and consequentialist ones, and proposes the need for agreement on minimal objectives about the evils to be avoided in genetic screening.)

Handyside, A. H., et al. "Birth of a Normal Girl after In Vitro Fertilization and Preimplantation Diagnostic Testing for Cystic Fibrosis." *New England Journal of Medicine* 327: 905–909 (1992).

Hayden, Michael R. "Predictive Testing for Huntington Disease: Are We Ready for Widespread Community Implementation?" *American Journal of Medical Genetics* 40: 515–517 (1991).

Hayes, Catherine V. "Genetic Testing for Huntington's Disease—A Family Issue." *New England Journal of Medicine* 327(20): 1449–1451 (November 12, 1992).

The Huntington's Disease Collaborative Research Group. "A Novel Gene Containing a Trinucleotide Repeat That Is Expanded and Unstable on Huntington's Disease Chromosomes." *Cell* 72: 971–983 (1993).

Kevles, Daniel J. "Social and Ethical Issues in the Human Genome Project." *National Forum* 73: 18–21 (1993).

Kolata, Gina. "Genetic Screening Raises Questions for Employers and Insurers." *Science* 232: 317–319 (April 18, 1986). (Discusses the range of issues raised by genetic screening, including testing young children, the privacy interests of employees, and the economic interests of employers and insurers to have genetic information. Provides background on the development of genetic screening technology in the workplace.)

Meissen, Gregory H., et al. "Understanding the Decision To Take the Predictive Test for

Huntington Disease." *American Journal of Medical Genetics* 39(4): 404–410 (June 15, 1991).

Nova. "Confronting the Killer Gene." Northbrook, IL: Coronet Film and Video, 1989. (Videotape.) (Examines the implications of a new laboratory test that can detect the presence of Huntington disease, a hereditary disorder characterized by nervous deterioration and certain death. Focuses on four individuals, including well-known musician Arlo Guthrie, who are all faced with a decision: to test or not to test.)

———. "Decoding the Book of Life." Northbrook, IL: Coronet Film and Video, 1989. (Videotape.) (Portrays the history of the Human Genome Project, placed in the context of the eugenics movement, development of genetic screening, treatment of genetic disease, and the choices that will need to be made regarding the use of molecular genetic techniques in human "breeding.")

———. "Genetic Gamble." Northbrook, IL: Coronet Film and Video, 1985. (Videotape.) (Explores the technical and ethical problems of applying breakthroughs in genetic engineering to the curing of inherited human disease. Examines research aimed at altering an individual's genetic code through insertion of gene sequences for missing enzymes into somatic cells. Illustrates the serious moral and ethical concerns raised by human genetics experimentation.)

Reilly, Philip. "Opinion: Advantages of Genetic Testing Outweigh Arguments against Widespread Screening." *Scientist* (January 21, 1991): 9, 11. (Discusses how the history of genetic testing in the United States for sickle-cell anemia and Tay-Sachs disease influences opinions about testing for cystic fibrosis [CF]. Reilly argues that as the accuracy rate approaches 90 percent, the advantages of testing even those persons without a family history of CF outweighs the negative aspects.)

Robertson, John A. "Legal Issues in Genetic Testing." In *The Genome, Ethics and the Law: Issues in Genetic Testing*, edited by the American Association for the Advancement of Science–American Bar Association National Conference of Lawyers and Scientists and the American Association for the Advancement of Science Committee on Scientific Freedom and Responsibility, pp. 79–110. Washington, DC: American Association for the Advancement of Science, 1992.

Shaw, Margery. "Invited Editorial Comment: Testing for the Huntington Gene: A Right to Know, a Right Not to Know, or a Duty to Know." *American Journal of Medical Genetics* 26: 243–246 (1987). (This brief article succinctly discusses the ethical problems surrounding patients' rights to obtain and to refuse knowledge after presymptomatic testing for Huntington disease.)

U.S. Office of Technology Assessment (OTA). *The Role of Genetic Testing in the Prevention of Occupational Disease*. Washington, DC: Office of Technology Assessment, 1983. (Chapter 9, "Application of Ethical Principles to Genetic Testing," discusses ethical questions relating to genetic testing in the workplace, including whether employers have duties to employees who may be at increased risk because of their genetic makeup or exposure to hazardous substances; whether genetic screening and monitoring in the workplace is ethical; whether workplace screening programs must be voluntary; and the actions that employers can ethically take using information obtained through genetic screening. Also provides background on ethical principles of autonomy, nonmaleficence, beneficence, and justice.)

Wertz, Dorothy. "Opinion: Lives in the Balance: Assessing the Risks of Waiting for Perfectly Accurate Tests." *Scientist* (January 21, 1991): 9, 11. (Argues against widespread testing for CF until the prediction rate improves and tests become widely available to all who need them, and asserts that pilot studies are needed to assess popular opinions about CF and carriers of CF, how they deal with risks, and how they can best be educated.)

Wexler, Nancy. "Presymptomatic Testing for Huntington's Disease: Harbinger of the New Genetics." *National Forum* 73: 22–26 (1993).

Wiggins, Sandi, et al. "The Psychological Consequences of Predictive Testing for Huntington's Disease." *New England Journal of Medicine* 327(20): 1402–1405 (November 12, 1992).

LONGER REFERENCES AND REFERENCES OF SECONDARY INTEREST

American Association for the Advancement of Science. *The Genome, Ethics and the Law*. Washington, DC: American Association for the Advancement of Science, 1992. (The first chapter, by Tabitha Powledge, "Ethical and Legal Implications of Genetic Testing: A Synthesis," provides background on genetic research and testing, control and use of genetic information in the workplace and in medicine, and the implications of testing for insurance, medical personnel, and commercial reasons. The chapter by Thomas Murray, "The Human Genome Project and Genetic Testing: Ethical Implications," discusses general issues that genetic testing presents for health care and research; using genetic tests to identify carriers, test newborns, neonates, and those at risk; workplace genetic testing; and the implications of testing for insurance and law enforcement purposes.)

Dorozynski, Alexander. "Privacy Rules Blindside French Glaucoma Effort." *Science* 252: 369–370 (April 19, 1991). (Report on the ethical implications of a plan in a French village to notify potential carriers of a genetic defect that can lead to blindness.)

Dworkin, Roger. "The New Genetics." In *BIOLAW: A Legal and Ethical Reporter on Medicine, Health Care and Bioengineering*, pp. 89–112. New York: University Publications of America, 1986. (Provides an overview of the legal and ethical issues involved in genetic screening, including the physician's obligations in genetic counseling, prenatal diagnosis, and conveying information to the patient and others; state-sponsored and mandatory employer screening; and genetic engineering and gene therapy.)

Faden, Ruth, et al. "A Survey to Evaluate Parental Consent As Public Policy for Neonatal Screening." *American Journal of Public Health* 72: 1347–1352 (1982). (Reports on a survey evaluating the impact of a Maryland requirement of parental consent for PKU screening of newborns.)

Grady, Denise. "The Ticking of a Time Bomb in the Genes." *Discover* 8: 26–35 (June 1987). (Excellent general discussion of testing and notification for Huntington disease, raising ethical issues of confidentiality and the right to know test results.)

Hilton, Bruce, et al., eds. *Ethical Issues in Human Genetics*. New York: Plenum Press, 1973. (Although this volume is now a bit dated, it contains articles discussing a number of still-timely issues, including the ethical uses of genetic knowledge, the ethics of genetic counseling [see Gustafson, in shorter references above], ethical issues in screening adults for Tay-Sachs disease, the ethics of genetic screening, implications of prenatal diagnosis for the human right to life and quality of life, privacy and genetic information, and the moral dimensions of parental decision making in genetic counseling.)

Holtzman, Neil. *Proceed with Caution: Predicting Genetic Risks in the Recombinant DNA Era*. Baltimore: Johns Hopkins University Press, 1989. (Discusses the implications of genetic research, testing, and screening for the medical profession [the author's primary intended audience]. Chapter 5 is devoted to genetic testing.)

Hunt, Morton. "The Total Gene Screen." *New York Times Magazine* (January 19, 1986): 33. (Provides a historical look at the development of genetic screening programs in the workplace, the protests waged against them, and the moral dilemmas they raise, including equal opportunity versus health protection, equal opportunity versus free enterprise, fairness to disabled versus fairness to the society at large, individual freedom versus social control, and knowledge versus privacy [or paternalism versus autonomy].)

Kevles, Daniel J., and Hood, Leroy. *The Code of Codes: Scientific and Social Issues in the Human Genome Project.* Cambridge, MA: Harvard University Press, 1992.

Lewis, Ricki. "Genetic Marker Testing: Are We Ready for It?" *Issues in Science and Technology* 4(1): 76–82 (1987). (Excellent discussion of some ethical issues surrounding genetic-marker testing, including confidentiality, privacy, public welfare, and the rights of minors and employers.)

Murray, Thomas. "Genetic Testing at Work: How Should It Be Used?" *Technology Review* 88(4): 51–59 (May/June 1985). (Provides an in-depth discussion of the moral issues raised in genetic testing at work. Distinguishes between genetic screening and monitoring, summarizes significant screening programs that have been instituted for PKU and sickle-cell anemia, the purposes of genetic screening and monitoring in the workplace, and the legal implications of screening programs.)

U.S. Department of Health and Human Services. Public Health Service. National Institutes of Health. Office for Protection from Research Risks. *Protecting Human Research Subjects: Institutional Review Board Guidebook.* Washington, DC: Government Printing Office, 1993. (Chapter 5 provides extensive discussion of ethical issues involved in genetic research. Other chapters contain in-depth discussions of such concepts as risk/benefit, informed consent, and privacy and confidentiality.)

U.S. Office of Technology Assessment. *Genetic Monitoring and Screening in the Workplace.* Washington, DC: Government Printing Office, 1990. (Chapter 1 summarizes ethical issues involved in workplace genetic screening. Chapter 7 considers such ethical issues as the voluntariness of workplace screening, access to information obtained from screening, communication and interpretation of test results, and criteria for monitoring workplace screening programs.)

U.S. President's Commission for the Study of Ethical Problems in Medicine and Biomedical and Behavioral Research. *Screening and Counseling for Genetic Conditions.* Washington, DC: Government Printing Office, 1983. (The Commission's conclusions—that genetic screening and counseling should be conducted in compliance with the five principles of confidentiality, autonomy, knowledge, well-being, and equity—are discussed briefly in the introduction and in depth in chapter 2, "Ethical and Legal Implications.")

Veatch, Robert. *Medical Ethics.* Boston: Jones and Bartlett, 1989. (Chapter 8, "Genetics and Reproductive Technologies," provides an introduction to genetic screening and testing, technical and legal background on prenatal screening and diagnosis, carrier testing and screening, and gene mapping and sequencing of the human genome. Ethical and public policy issues covered include freedom versus coercion, confidentiality versus disclosure, access to genetic testing services, probable benefits and harms, human gene therapy, and genetic engineering.)

Wertz, Dorothy, and Fletcher, John, eds. *Ethics and Human Genetics: A Cross-Cultural*

Perspective. Berlin: Springer-Verlag, 1989. (Especially relevant are chapter 2.19.2, "Ethical Problems", which discusses genetic counseling, prenatal diagnosis, and genetic screening; chapter 2.19.5, "Ethical Issues and Future Trends"; and chapter 3, "Ethics and Human Genetics: A Cross-Cultural Perspective", which covers the major ethical problems in the practice of human genetics in contemporary society, the need to develop approaches to educate and formulate resolutions for these problems, and the responsibilities of geneticists to formulate public policies on these problems.)

——. "An International Survey of Attitudes of Medical Geneticists toward Mass Screening and Access to Results." *Public Health Reports* 35(10): 104 (January/February 1989). (Discusses genetic researchers' attitudes about patients' rights to information and to confidentiality about the results of screening tests for genetic diseases.)

3. *Chapter 8, Ethics and Eugenics*

SHORTER REFERENCES OF PRIMARY INTEREST

Bishop, Jerry E. "Unnatural Selection." *National Forum* 73: 27–29 (1993).

Carlson, Elof. "R. L. Dugdale and the Jukes Family: A Historical Injustice Corrected." *BioScience* 30: 535–539 (1980). (Describes how Dugdale's original work on the Jukes family—the cultural cousins of the Kallikaks in the pro-eugenics movement—was twisted by eugenicists from its environmental explanation of the family's disabilities to a hereditary one.)

Editorial. "The Jukes Redeemed." *Scientific American* 243: 88–89 (November 1980). (Discusses the distorted evidence of feeblemindedness in the Jukes family and the origination of the distortions with subsequent editors, rather than with Dugdale, the original compiler of the case study.)

Fancher, Raymond. "Henry Goddard and the Kallikak Family Photographs." *American Psychologist* 47: 589–590 (1987). (Discusses accusations that Goddard retouched photographs of subjects in his study of the feebleminded in order to support his findings, suggests that such manipulation of evidence may be innocent, and considers some of the difficulties in writing accurate and fair history.)

Gould, Stephen J. "Carrie Buck's Daughter." *Natural History* 93: 14–18 (1984). (Discusses the influence of the eugenics movement on immigration restrictions and involuntary sterilization, focusing on the *Buck v. Bell* case. Discloses that Carrie Buck was institutionalized not because she was feebleminded but because she had been raped and became pregnant out of wedlock [and subsequently gave birth to an illegitimate daughter falsely diagnosed as feebleminded as well].)

O'Brien, Margaret, and Levine, Carol. "The XYY Controversy: Researching Violence and Genetics." *Hastings Center Report* (Special Supplement, August 1980), pp. 1–31. (Includes excerpts from two conferences on genetic determinants, violent behavior, and the ethics of research and research review, stemming from controversial research exploring whether men with an extra Y chromosome are more prone to violent behavior. Raises issues of researcher responsibility to subjects [especially regarding informed consent], their families, and society; sloppy research; and media responsibilities regarding reporting of controversial research.)

Paul, Diane. "Books: A History of the Eugenics Movement and of Its Multiple Effects on Public Policy." *Scientific American* 254: 27–31 (January 1986). (Book Review of Daniel Kevles, *In the Name of Eugenics: Genetics and the Uses of Human Heredity* [see below]. Generally positive review of Kevles's historical treatment of eugenics, particularly his consideration of how that history can inform current policy makers about the social benefits and pitfalls of individual reproductive decisions. The review discusses the difficulty of defining "eugenics" and criticizes both the appropriateness of Kevles's sharp distinction between "old" and "newer" eugenics movements and his portrayal of the historical figures involved as "good guys" and "bad guys.")

Wexler, Nancy. "Presymptomatic Testing for Huntington's Disease: Harbinger of the New Genetics." *National Forum* 73: 22–26 (1993).

LONGER REFERENCES AND REFERENCES OF SECONDARY INTEREST

Davenport, Charles B. *Medical Genetics and Eugenics*. Philadelphia: Women's Medical College of Pennsylvania, 1940.

——. *Heredity in Relation to Eugenics*. New York: Henry Holt, 1911. (Written to promote interest in the study of heredity and the eugenics movement toward the end of reducing the reproduction of "defectives and delinquents" and controlling immigration.)

Goddard, Henry H. *Feeble-Mindedness: Its Causes and Consequences*. New York: Macmillan, 1914. (Describes itself as a report on 327 cases of "feeble-mindedness" studied at the Training School for Feeble-minded Girls and Boys in Vineland, New Jersey.)

——. *The Kallikak Family: A Study in the Heredity of Feeble-Mindedness*. New York: Macmillan, 1912; reprinted by Arno Press, New York, 1973. (Written by the director of the research laboratory of the Vineland Training School. Based on field work conducted at the family homes of Vineland's resident children. Concludes that about 65 percent of the families have the "hereditary trait" for feeblemindedness. The final chapter, "What Is to Be Done?" considers segregation and colonization of the feebleminded and concludes that sterilization is a necessary temporary measure.)

Haldane, J. B. S. *Heredity and Politics*. New York: Norton, 1938. (Argued that current knowledge of heredity was an inadequate basis to justify much of the legislation authorizing sterilization and restricting immigration supposedly based on it.)

Kevles, Daniel. *In the Name of Eugenics: Genetics and the Uses of Human Heredity*. New York: Knopf, 1985. (Kevles's history of the eugenics movement shows how early notions of genetic perfectability appeared in the United States; how the movement was embraced by political liberals as well as conservatives, feminists as well as sexists, and was used to support a wide variety of social policy goals; and how public support for eugenics programs waned in the wake of Nazism. Kevles concludes that subsequent developments, such as population control and medical genetics, should be distinguished from the earlier eugenics movement.)

——, and Hood, Leroy, eds. *The Code of Codes: Scientific and Social Issues in the Human Genome Project*. Cambridge, MA: Harvard University Press, 1992. (Several chapters relate to the connection between eugenics and modern technologies in human genetics. See especially chapters 1, 13, and 14.)

Newman, Horatio H. *Readings in Evolution, Genetics, and Eugenics*. Chicago: University of Chicago Press, 1921; reprinted 1969. (Part five, on eugenics, contains materials on the restriction of undesirable germplasm, control of immigration, more discriminating marriage laws, segregation of defectives, and drastic measures.)

Race Betterment Society. *Proceedings of the First National Conference on Race Betterment.* Battle Creek, MI: Race Betterment Society, 1914. (Participants, including Davenport [see entries above], discuss the relationship of eugenics and immigration to race betterment, among other topics, at a conference convened to assemble "evidence as to the extent to which degenerative tendencies are actively at work in America, and to promote agencies for Race Betterment.")

Santos, M. A. *Genetics and Man's Future: Legal, Social and Moral Implications of Genetic Engineering.* Springfield, IL: Charles C. Thomas, 1981. (Focuses on genetics, law, and human improvement. Argues that children have the right to be born free of genetic defects and that parents thus have the right to genetic information that will enable them to produce such children. Advocates a limited program of eugenics to prevent needless human suffering. The proposed limitations comprise mechanisms designed to prevent the abuses of prior eugenics programs.)

Smith, J. David. *Minds Made Feeble: The Myth and Legacy of the Kallikaks.* Rockville, MD: Aspen Systems Corp., 1985. (Uncovers Goddard's errors in tracing the Kallikaks' "feeble-mindedness" to hereditary factors and shows how easily his conclusions were adopted by groups as the basis for political agendas that deny opportunities to certain segments of society. Questions Goddard's classification of Deborah Kallikak, and thus her family, as feebleminded and explores the factors that influenced Goddard's findings of a genetic explanation for mental retardation. Describes Goddard's influence on limiting immigration and on the Nazi racial hygiene program and notes that echoes of Goddard's conclusions—that ignorance, poverty, and social pathology have a genetic basis—continue to be influential.)

Trombley, Stephen. *The Right to Reproduce: A History of Coercive Sterilization.* London: Weidenfeld & Nicholson, 1988. (Takes the view that compulsory and coercive sterilization are gross violations of a basic human right and that most policies of eugenic sterilization are racist, sexist, and imperialist. Asserts that the history of the eugenics debate has shown that sterilization of the "unfit" was unlikely to solve either social or medical problems.)

Woodside, Moya. *Sterilization in North Carolina: A Sociological and Psychological Study.* Chapel Hill: University of North Carolina Press, 1950. (Study of North Carolina's eugenics program of voluntary sterilization "in selected cases of normal or borderline intelligence." Recommends education programs to promote widespread acceptance, especially "among the lower-class Negro groups, since it is here that fertility is highest and mental defect more prevalent.")

Psychology

1. *Chapter 9, Ethical Issues in Animal Experimentation*

SHORTER REFERENCES OF PRIMARY INTEREST

Positions in Support of Less-Restricted Use of Laboratory Animals in Research

Cohen, Carl. "The Case for the Use of Animals in Biomedical Research." *New England Journal of Medicine* 315(14): 865–869 (October 1986). (Argues that animals have no rights and that "speciesism" is an inadequate rationale for prohibiting the use of animals in research. Concludes that alternatives should be utilized where possible but that animal research cannot be eliminated; that the use of animals in research should be increased rather than decreased to avoid using humans as subjects; and that the perspectives of opponents to animal experimentation are seldom consistent.)

Loeb, Jerod, et al. "Human vs. Animal Rights: In Defense of Animal Research." *Journal of the American Medicine Association* 262(19): 2716–2720 (1989). (Condensed version of the American Medical Association's detailed analysis of the controversy over the use of animals in research, including reports on the destructive and violent actions of animal-rights activist movements and discussion of the consequences for research if the activists prevail.)

Miller, Neal. "The Value of Behavioral Research on Animals." *American Psychologist* 40(4): 423–440 (April 1985). (Refutes the position of some animal-rights activists that animal experimentation has been valueless by discussing the many varied contributions that animal experimentation has made to medicine and health, including benefits to animals, protection of people and crops, behavior therapy and medicine, stress and pain therapy, and learning and memory deficits.)

Positions Sympathetic to Animal Rights

Singer, Peter. "The Significance of Animal Suffering." *Behavioral and Brain Sciences* 13(1): 9–12 (March 1990). (Argues that animals suffer, that their suffering is of moral consequence, and that there is no characteristic that makes humans universally more morally significant than animals. Concludes that "speciesism"—the preferential treatment of humans over animals—is not justified and that we should instead consider the suffering of animals as on a moral par with our own.)

——. "Tools for Research . . . or What the Public Doesn't Know It Is Paying For." In *Animal Liberation: A New Ethics for Our Treatment of Animals*, pp. 27–91. New York: Avon Books, 1975. (Argues for the abolition of animal research. Uses the principle of "equal consideration of interests"—that all sentient beings have the same stake in their own existence—to argue that giving primacy to the interests of one species over another on any basis other than its capacity to suffer is illegitimate.)

Zak, Steven. "Ethics and Animals." *Atlantic Monthly* 263(3): 68–74 (March 1989). (Discusses the "radical" views of contemporary animal-rights groups relative to the past; considers differences between virtue and rights approaches to the ethical treatment of animals, showing how the former inadequately protects animals; and concludes that legal recognition of animals' rights is necessary to their adequate protection.)

Other Sources on the Use and Treatment of Animals in Research

American Psychological Association. *Casebook on Ethical Principles of Psychologists.* Washington, DC: American Psychological Association, 1987. (See Principle 10, "Care and Use of Animals." A one-page statement of standards for the care and treatment of animals.)

Donnelly, Strachan, and Nolan, Kathleen. "Animals, Science, and Ethics." *Hastings Center Report* 20(3): 1–32 (Special Supplement May/June 1990). (An informative, in-depth discussion of the ethical implications of the use of animals in research. Includes an introductory overview; summary of applicable ethical theories and their usefulness applied to animals; consideration of rationales used to justify using animals in science and how best to determine animal suffering; overview of animal care and use committees and review procedures; and discussion of policy issues relating to the use of animals in research, testing, and education.)

Herzog, Harold. "Comment: Conflicts of Interests: Kittens and Boa Constrictors, Pets and Research." *American Psychologist* 45(3): 246–248 (March 1991). (Argues that even seemingly benign relations with animals, such as keeping carnivorous animals as pets, raises some of the same questions as the use of animals in scientific research.)

———. "Discussing Animal Rights and Animal Research in the Classroom." *Teaching of Psychology* 17: 90–94 (1990). (Reviews utilitarian and rights arguments for animal liberation, presents a classroom exercise involving four cases which students decide as members of an animal care and use committee, and reports that students' participating in the exercise raised their awareness of the issue and its complexity.)

———. "The Moral Status of Mice." *American Psychologist* 43(6): 473–474 (June 1988). (This short but poignant article develops a typology of mice to demonstrate how the same animals given different roles and labels in different settings receive different moral consideration, suggesting that the labels animals receive influence their moral status and treatment.)

Sigma Xi, the Scientific Research Society. "Statement on the Use of Animals in Research." *American Scientist* 80: 73–76 (January/February 1992). (Following a request for commentary from all chapters and clubs of Sigma Xi, this statement reflects the organization's policy on the responsible use of animals in research, based on the assumption that "well-conducted research with animals has provided, and continues to provide, information, ideas, and applications that can be obtained in no other way." Proposes limiting research using animals to that which is necessary, finding alternatives where feasible, and ensuring that animals do not suffer unnecessarily.)

U.S. Department of Agriculture. "Animal Welfare Rules and Regulations." *Code of Federal Regulations* Title 9, Part 1, "Definition of Terms;" Part 2, "Regulations;" and Part 3 "Animal and Plant Health Inspection Service."

LONGER REFERENCES AND REFERENCES OF SECONDARY INTEREST

Positions in Support of Less-Restricted Use of Laboratory Animals in Research

Feeney, Dennis. "Human Rights and Animal Welfare." *American Psychologist* 47(6): 593–597 (June 1987). (A defense of animal experimentation written by a researcher crippled by incurable paraplegia. Criticizes the antivivisectionist movement for neglecting the suffering of humans caused by failure to experiment on animals.)

Fox, Michael. *The Case for Animal Experimentation: An Evolutionary and Ethics Perspective.* Berkeley: University of California Press, 1986. (A defense of experimentation with animals in scientific research. Includes a history of research with animals, describes the development of the antivivisectionist movement, and explores the moral status of animals in relation to humans. Concludes that animal welfare is a lower concern than human welfare, although animals should not be mistreated and their suffering should be avoided where possible.)

McCabe, Katie. "Who Will Live, Who Will Die?" *Washingtonian* 21(11): 112–118, 153–157 (August 1986). (An in-depth look at the effect of the animal-rights movement on experimental research using animals, detailing lifesaving treatments for persons, the development of which depends on animal experimentation, as well as the tactics used by animal-rights groups protesting the use of animals in research.)

National Academy of Sciences and The Institute of Medicine. Committee on the Use of Animals in Research. *Science, Medicine, and Animals.* Washington, DC: National Academy Press, 1991. (A pamphlet defending the use of animals in research on the basis of the important scientific and medical advances that have depended on it, defending current research practices involving animals in the United States today as one in which "mistakes are rare," and criticizing the animal-rights movement as attempting to eliminate all research using animals.)

Nozick, Robert. "About Mammals and People." *New York Times Book Review* (November 27, 1983), p. 11. (In this review of Tom Regan's *The Case for Animal Rights*, Nozick opposes Regan's argument that animals have rights entitled to be protected, his claims about the level of animals' psychological and emotional sophistication and the low status he accords to severely mentally disabled humans, but agrees with the necessity of formulating a balanced and ethically sensitive position on the moral status of animals.)

Pardes, Herbert; West, Anne; and Pincus, Harold. "Physicians and the Animal-Rights Movement." *New England Journal of Medicine* 324(4): 1640–1643 (June 6, 1991). (Warns of the dangers that animal-rights activists pose to medical research; discusses the tactics of activists, including frightening the public, using "psychological warfare," violence, legal maneuvers, and the myth of alternatives; and urges physicians to inform their patients about the benefits and importance of experimentation with animals.)

Positions Sympathetic to Animal Rights

Forum. "Just Like Us." *Harper's* 277(5): 43–52 (August 1988). (Debate on animal rights with pro-rights activists Gary Francione, professor at University of Pennsylvania Law School, and Ingrid Newkirk, national director of People for the Ethical Treatment of Animals [PETA], and with Arthur Caplan, director of the Center for Biomedical Ethics at the University of Minnesota, and Roger Goldman, constitutional law professor at the St. Louis University School of Law.)

Regan, Tom. *The Case for Animal Rights.* Berkeley: University of California Press, 1983. (Argues that animals have rights: valid claims against others to particular treatment and assistance in achieving that treatment. The existence of these rights, he contends, requires extensive changes in the way humans treat animals, particularly in the areas of meat eating and animal experimentation.)

Singer, Peter. *Animal Liberation: A New Ethics for Our Treatment of Animals.* New York: Avon Books, 1975. (See also the revised edition, published in 1990.)

Other Sources on Animal Experimentation

Clingerman, Karen. "Ethical and Moral Issues Relating to Animals 1979–1988." *Quick Bibliography Series.* Beltsville, MD: U.S. Department of Agriculture, 1988). (Contains short annotations of 270 citations.)

Dewsbury, Donald. "Early Interactions between Animal Psychologists and Animal Activists and the Founding of the APA Committee on Precautions in Animal Experimentation." *American Psychologist* (March 1990): 315–327. (Demonstrates that the contemporary conflict between animal psychologists is not new but dates back to the Victorian antivivisectionist movement, and discusses the effects of the early controversy.)

Gallup, Gordon, and Beckstead, Jason. "Attitudes toward Animal Research." *American Psychologist* 43(6): 474–476 (June 1988). (Summarizes results of a survey of 263 undergraduate students. The survey revealed that 76 percent were concerned about pain and suffering in animals, yet 81 percent preferred to see animals used in research rather than see humans die or suffer from disease, 66.9 percent agreed that drugs and surgical procedures should be tested on animals before being administered to humans, and 62 percent believed that some animal research is necessary.)

Jasper, James M., and Nelkin, Dorothy. *The Animal Rights Crusade: The Growth of a Moral Protest.* NY: Maxwell Macmillan International, 1992. (Traces the history of the animal-rights movement.)

Mitchell, Graham. "Guarding the Middle Ground: The Ethics of Experiments on Animals." *Current Contents* 22: 6–13 (1990). (Argues that use of ethics committees is the best means to mediate between the conflicting interests of researchers desiring to use animals in research and antivivisectionists.)

Newman, Alan. "Research versus Animal Rights: Is There a Middle Ground?" *American Scientist* 77: 135–137 (March-April 1989). (Reports on the rise of animal-rights activism and its impact on researchers, including the development of alternatives to experimental use of animals.)

U.S. Department of Health and Human Services. Public Health Service. National Institutes of Health. Office for the Protection from Research Risks. *Public Health Service Policy on Humane Care and Use of Laboratory Animals.* Washington, DC: Government Printing Office, 1989. (Comprehensive, detailed description of the Public Health Service regulations regarding the care and treatment of animals used in scientific research.)

——— and Applied Research Ethics National Association. *Institutional Animal Care and Use Committee Guidebook.* Washington, DC: U.S. Government Printing Office, NIH Publication No. 92–3415 (looseleaf).

Ulrich, Roger. "Commentary: Animal Rights, Animal Wrongs, and the Question of Balance." *Psychological Science* 2(3): 197–201 (May 1991). (Argues for balance in the use and treatment of animals, contending that both the antivivisectionists and the researchers who conduct experimentation with animals are guilty of excesses.)

2. *Chapter 10, Research Involving Human Subjects; Chapter 11, Research Involving Human Subjects: The Administration of Alcohol; Chapter 12, The Ethics of Deception in Research*

SHORTER REFERENCES OF PRIMARY INTEREST

American Psychological Association. *Casebook on Ethical Principles of Psychologists.* Washington, DC: American Psychological Association, 1987. (Sets forth Principle 9, "Research with Human Participants," which is covered much more extensively in the APA's *Ethical Principles*, cited below in the section containing longer references.)

Baumrind, Diana. "Research Using Intentional Deception." *American Psychologist* 40(2): 165–174 (February 1985). (Reviews ethical issues concerning intentional deception ten years after the publication of APA guidelines on research with human subjects. Concludes that the guidelines have not reduced intentional deception. Offers alternative research strategies for avoiding deception and for debriefing subjects.)

Capron, Alexander. "Is Consent Always Necessary in Social Science Research?" In *Ethical Issues in Social Science Research*, edited by Tom Beauchamp, Ruth Faden, R. Jay Wallace, and LeRoy Walters, pp. 215–231. Baltimore: Johns Hopkins University Press, 1982. (Argues that informed consent may not be necessary in social science research when there are other means to fulfill its underlying purposes, including the promotion of autonomy and the protection of privacy.)

Dane, Francis. *Research Methods*. Belmont, CA: Brooks Cole, 1990. (Chapter 3, "Research Ethics," provides a good general discussion of unethical conduct that takes place before, during, and following the research project, including lack of informed consent, deception, physical and psychological harm, and the placing of limits on the autonomy of subjects. Also dealt with are problems of debriefing, anonymity, and confidentiality, as well as relevant federal regulations and the APA's code of ethical conduct.)

Dworkin, Gerald. "Must Subjects Be Objects?" In *Ethical Issues in Social Science Research*, edited by Beauchamp et al., pp. 246–254. (Challenges the assumption that the ethical principle that calls for treating people as "ends" rather than as "means to ends" is self-evident and unproblematic.)

Elms, Alan. "Keeping Deception Honest: Justifying Conditions for Social Scientific Research Strategems." In *Ethical Issues in Social Science Research*, edited by Beauchamp et al., pp. 232–245. (Argues that deception in social scientific research is ethically tolerable if the following conditions are met: no alternative to deception exists, benefits outweigh harms, subjects are free to withdraw at any time without penalty, any harm that might be caused is temporary, and subjects are fully debriefed following the research.)

Erickson, Kai. "A Comment on Disguised Observation in Sociology." *Social Problems* 14: 366–373 (1967). (Discusses the ethical problems associated with disguised participant observation research, including its potential for harming subjects, the reputation of the profession, and students, as well as the accuracy and integrity of science.)

Garfield, Sol. "Ethical Issues in Research on Psychotherapy." *Counseling and Values* 31(2): 115–125 (April 1987). (Discusses differences in attitudes about treatment of human subjects in research versus therapy, the ethical problems associated with research designs that include nontreatment control groups, and proposed solutions to these problems.)

Kelman, Herbert C. "Privacy and Research with Human Beings." *Journal of Social Issues* 33: 169–195 (1977). (Discusses the threat to the subject's privacy resulting from social and psychological research, ways to counteract such threats, and the conditions under which invasion of privacy may be justified.)

Macklin, Ruth. "The Problem of Adequate Disclosure in Social Science Research." In *Ethical Issues in Social Science Research*, edited by Beauchamp et al., pp. 193–214. (Supports the application of federal regulations for the protection of human subjects to the social sciences as well as the medical sciences, arguing that the dangers of harm to human subjects and the need for informed consent make the principles underlying such regulations, including respect for persons and a utilitarian weighing of social benefits against personal costs, equally applicable. Concludes that outright deception, withholding information relevant to the subject's decision to participate, and disguised participant observation are unethical.)

May, William. "Doing Ethics: The Bearing of Ethical Theories on Fieldwork." *Social*

Problems 27: 358–370 (1980). (Discusses the relevance of teleological, utilitarian, deontological, critical philosophical, and conventional ethical theories to fieldwork.)

Milgram, Stanley. *Obedience to Authority: An Experimental View.* New York: Harper & Row, 1974. (In "Problems of Ethics in Research" [pp. 193–202], Milgram defends the ethical legitimacy of his experiments that involved the deception of human subjects. Milgram's scientific reports on his research and reactions to them are listed in the section below containing longer and secondary references.)

——. "Issues in the Study of Obedience: A Reply to Baumrind." *American Psychologist* 19: 848–852 (1964). (Milgram responds to criticism of his studies.)

Nuremberg Code. In *Trials of War Criminals before the Nuremberg Military Tribunals under Control Council Law No. 10*, vol. 2, pp. 181–182. Washington, D.C.: U.S. Government Printing Office, 1949. Also reprinted in Robert J. Levine, *Ethics and Regulation of Clinical Research*, 2d ed., pp. 425–426. Baltimore: Urban and Schwarzenberg, 1986. (Drafted in response to the atrocities committed in Nazi experimentation with human subjects. Sets forth the basic principles for ethical experimentation, including informed consent, social benefit that outweighs risk to subjects, avoidance of harm, necessity, qualified researchers, and the subject's right to terminate the experiment at any point.)

U.S. National Commission for the Protection of Human Subjects of Biomedical and Behavioral Research. *Belmont Report: Ethical Principles and Guidelines for the Protection of Human Subjects of Biomedical and Behavioral Research.* Washington, DC: U.S. Government Printing Office, 1978. DHEW Publication No. (OS) 78-0012. Reprinted in *Federal Register* 44: 23192 (April 18, 1979). (Summarizes and discusses the basic ethical principles identified by the National Commission for the Protection of Human Subjects of Research: respect for persons, beneficence, and justice.)

Warwick, Donald P. "Deceptive Research: Social Scientists Ought to Stop Lying." *Psychology Today* (February 1975): 28–39, 105–106. (Summarizes a number of social science research studies that involved deliberate deception of subjects, including a number by Stanley Milgram, and argues that they are unethical.)

World Medical Association. "Declaration of Helsinki." As amended by the 41st World Medical Assembly, Hong Kong, September 1989. Reprinted in *Law, Medicine and Health Care* 19(3–4): 264–265 (Fall/Winter 1991).

LONGER REFERENCES AND REFERENCES OF SECONDARY INTEREST

American Psychological Association. *Ethical Principles in the Conduct of Research with Human Participants.* Washington, DC: American Psychological Association, 1982. (Based on APA Principle 9, "Research with Human Participants." Discusses the ethical dilemmas involved in conducting psychological research involving human subjects and the guidelines that should shape their resolution, including determining whether to undertake a particular research project involving human subjects, informed consent, when if ever deception of human subjects is acceptable, what constitutes harm to and exploitation or coercion of human subjects, and responsibilities of researchers for debriefing subjects at the conclusion of research.)

Beauchamp, Tom L. and Childress, James F. *Principles of Biomedical Ethics*, 3d ed. New York: Oxford University Press, 1989.

Beauchamp, Tom; Faden, Ruth; Wallace, R. Jay; and Walters, LeRoy, eds. *Ethical Issues in Social Science Research*. Baltimore: Johns Hopkins University Press, 1982. (Includes chapters on the foundations of ethics in social science research, harms and benefits, informed consent, and deception.)

Bower, Robert T., and Gasparis, Priscilla. *Ethics in Social Research: Protecting the Interests of Human Subjects*. New York: Praeger, 1978. (Good historical background on the regulation of research involving human subjects. Chapters also cover risks to subjects of social research, including deception, coercion, breach of privacy and confidentiality, stress and risks to the larger society, informed consent, balancing harms and benefits, and regulation.)

Brobeck, Sonja. "The Need for Better Ethical Guidelines for Conducting and Reporting Research." *Journal of Curriculum and Supervision* 5: 194–200 (Winter 1990). (Briefly summarizes the development of regulations establishing standards for the treatment of human subjects in research. Argues for improved standards in such areas as researcher qualifications, specification of the characteristics of the particular group to be studied, methods of data collection, maintenance of the subject's privacy, and anonymity of subjects.)

Budiansky, Stephen. "Playing Roulette with Experimental Drugs." *U.S. News and World Report* 103: 58–59 (July 13, 1987). (Reviews ethical issues arising from physician-researchers' zeal to enroll subjects in experimental drug tests that may not be in the subjects' best interests.)

Bulmer, Martin, ed. *Social Research Ethics: An Examination of the Merits of Covert Participant Observation*. New York: Holmes & Meier, 1982. (Introduces the ethical problem of covert participant observation. Describes five case studies, contains several essays discussing ethical theory as it relates to covert methods, and describes the pros and cons of covert participant observation.)

du Toit, Brian. "Ethics, Informed Consent, and Fieldwork." *Journal of Anthropological Research* 36(3): 274–286 (1980). (Discusses the importance of obtaining informed consent in fieldwork-type research.)

Forsyth, D. R. *An Introduction to Group Dynamics*, pp. 373–376. Monterey, CA: Brooks Cole, 1983. (Discusses the "robber's cave" experiments that divided twenty-two eleven-year-old boys at a summer camp into two groups and encouraged the groups to become increasingly aggressive toward one another.)

Freund, Paul A., ed. *Experimentation with Human Beings*. New York: George Braziller, 1969. (A classic text on research involving human subjects.)

Jonas, Hans. "Philosophical Reflections on Experimenting with Human Subjects." *Daedalus* 98(2): 219–247 (Spring 1969). (Considers the use of human subjects to be a balance of social needs versus matters of individual sacrosanctity and personal dignity. Argues that consent to experimentation, regardless of the importance of the research, is insufficient because of the primacy of the individual and the personal sacrifice involved. Also argues that it is necessary to consider how proximate the subject's interests are to those of the interests of the researcher and whether the experiment relates to the subject's own disease.)

Katz, Jay. *Experimentation with Human Beings*. New York: Russell Sage Foundation, 1972. (A classic text on research involving human subjects.)

Keith-Spiegel, Patricia, and Koocher, Gerald. *Ethics in Psychology: Professional Standards and Cases*. New York: Random House, 1985. (This text was designed both for teaching and as a consultation resource for use by individual psychologists. Of particular relevance are chapters 1, on codes of ethical conduct for psychologists and enforcement; 2, on the roles and functions of ethics committees; 13, on scholarly publishing and teaching; 14, on research issues, including discussions of the ethical treatment of subjects, assessing risks and benefits of research, deception and concealment, privacy, and confidentiality of research inside and outside the lab; and 15, "Psychology and the Public Trust.")

Levine, Robert J. *Ethics and Regulation of Clinical Research*, 2d ed. Baltimore: Urban and Schwarzenberg, 1986. (Excellent resource on research involving human subjects.)

Mendelson, Jack, and Mello, Nancy, eds. *The Diagnosis and Treatment of Alcoholism.* New York: McGraw-Hill, 1987. (Of particular relevance is chapter 1, "Diagnostic Criteria for Alcoholism.")

Milgram, Stanley. "Some Conditions of Obedience and Disobedience to Authority." *Human Relations* 18: 57–76 (1965). (This article and the article listed in the following entry describe the now-famous Milgram experiments, which involved the deception of human subjects in order to test their willingness to inflict pain on other human beings in obedience to authority.)

———. "Behavioral Study of Obedience." *Journal of Abnormal And Social Psychology* 6(4): 371–378 (1963).

Miller, Arthur G. *The Obedience Experiments: A Case Study of Controversies in Social Science*. New York: Praeger, 1986. (Draws together and reviews a diversity of social science critiques of the methodology and theory of deception used in Milgram's obedience research. Introduces Milgram's research program, reviews its extensions and replications in other laboratories, and discusses the ethics involved [chapter 5], criticisms of Milgram's method and alternatives to deception, and the implications of Milgram's work for understanding the Holocaust.)

Reece, Robert D., and Siegel, Harvey. *Studying People: A Primer in the Ethics of Social Research*. Macon, GA: Mercer University Press, 1986. (Discusses ethical problems in social research, including participant research, privacy and deception, informed consent, and harm. Uses a case-oriented approach.)

Shils, Edward. "Social Inquiry and the Autonomy of the Individual." In *Social Research Ethics: An Examination of the Merits of Covert Participant Observation*, edited by Martin Bulmer. New York: Holmes & Meier, 1982. (Discusses the invasion of privacy accompanying social scientific fieldwork; the vagueness of ethical problems raised by such intrusion, which include lack of respect for human dignity, autonomy, and privacy; the origins and importance of respect for privacy in our society as an aspect of individuality; and the necessity of respecting privacy in participant-observer research by requiring informed consent and prohibiting experimentation involving manipulation of subjects.)

Sieber, Joan, ed. *The Ethics of Social Research*. 2 vols. New York: Springer-Verlag, 1982.

Sieber, Joan E. *Planning Ethically Responsible Research: A Guide for Students and Internal Review Boards*. Applied Social Research Methods Series, vol. 31. Newbury Park, CA: Sage, 1992.

———, and Stanley, Barbara. "Ethical and Professional Dimensions of Socially Sensitive Research." *American Psychologist* 43(4): 49–55 (January 1988). (Proposes a taxonomy of analysis for considering the ethical dimensions of socially sensitive research, both in the formulation of the research question and in the conduct of the research and treatment of subjects. The taxonomy covers considerations of privacy, confidentiality, research methodology, deception, informed consent, justice and equitable treatment, scientific freedom, ownership of data, the values of epistemology of social scientists, and risk versus benefit.)

Smith, Steven, and Richardson, Deborah. "Amelioration of Deception and Harm in Psychological Research: The Important Role of Debriefing." *Journal of Personality and Social Psychology* 44(5): 1075–1082 (1983). (Reports on the effects of deception and harm on research participants in psychology experiments, based on the perceptions of 464 undergraduate students enrolled in psychology courses. Approximately 20 percent of the subjects reported harm, but there was no correlation between deception and perception of harm. The authors also consider the role of debriefing in eliminating negative effects for those participants who perceived they had been harmed, and conclude that participation as research subjects may be beneficial.)

U.S. Department of Health and Human Services. "Protection of Human Subjects." *Code of Federal Regulations*, Title 45, Part 46 (1990). (Sets forth federal regulations for the protection of human subjects—the institutional review board system.)

U.S. Department of Health and Human Services. Public Health Service. National Institutes of Health. Office for Protection from Research Risks. *Protecting Human Research Subjects: Institutional Review Board Guidebook*. Washington, DC: Government Printing Office, 1993. (Explicates the federal regulations for the protection of human research subjects. Provides extensive discussion of such concepts as risk/benefit, informed consent, and privacy and confidentiality. Discusses the human subjects issues raised in the context of various kinds of research [e.g., drug trials, AIDS research, genetic research] as well as special concerns necessitated by the involvement in research of certain classes of subjects [e.g., children, prisoners, cognitively impaired persons].)

Veatch, Robert M. *The Patient as Partner: A Theory of Human-Experimentation Ethics*. Bloomington: Indiana University Press, 1987. (Includes chapters discussing the ethics of research involving human subjects, covering such topics as justifications for research, informed consent, and emerging themes and controversies.)

——. *Medical Ethics*. Boston: Jones and Bartlett, 1989.

3. *Chapter 13, Misconduct in Science*

National Academy of Sciences. *Responsible Science: Ensuring the Integrity of the Research Process*. 2 vols. Washington, DC: National Academy Press, 1992.

National Academy of Sciences. *On Being a Scientist*. Washington, DC: National Academy Press, 1989.

Nova. "Do Scientists Cheat?" Northbrook, IL: Coronet Film and Video, 1988. (Videotape.)

Sigma Xi, the Scientific Research Society. *Honor in Science*. New Haven, CT: Sigma Xi, the Scientific Research Society, 1986.

Teich, Albert, and Frankel, Mark. *Good Science and Responsible Scientists: Meeting the Challenge of Fraud and Misconduct in Science*. Washington, DC: American Association for the Advancement of Science, 1992.

U.S. Department of Health and Human Services. Public Health Service. "Responsibilities of Awardee and Applicant Institutions for Dealing with and Reporting Possible Misconduct in Science." *Federal Register* 54(151): 32446–32451 (August 8, 1989).

U.S. Department of Health and Human Services. Public Health Service. Office of the Assistant Secretary for Health, Office of Health Planning and Evaluation, and Office of Scientific Integrity Review. *Data Management in Biomedical Research: Report of a Workshop* (April 1990).

History

1. *Chapter 16, The Historian's Code of Ethics; Chapter 17, The Use and Interpretation of Historical Documents; Chapter 21, Intellectual Property*

SHORTER REFERENCES OF PRIMARY INTEREST

Plagiarism

American Historical Association. "Statement on Plagiarism and Related Misuses of the Work of Other Authors." In *Statement on Standards of Professional Conduct*, pp. 13–16. Washington, DC: American Historical Association, 1993. (The history profession's official statement defining plagiarism and appropriate responses to its occurrence.)

Brookes, Gerry H. "Exploring Plagiarism in the Composition Classroom." *Freshman English News* 17: 31–35 (1989). (Offers a number of strategies for teaching about plagiarism, including assigning students to interview faculty and administrators about their experience in dealing with plagiarism and writing personal essays speculating about the causes of plagiarism.)

Clymer, Kenton. "Checking the Sources: John Hay and Spanish Possessions in the Pacific." *Historian* 48: 82–87 (November 1985). (An interesting article useful for demonstrating the issues that are raised by the use and interpretation of historical documents. Critiques an article on John Hay and argues that the conclusions drawn by its author are based on a misreading of a critical document.)

Hoyt, Michael. "Malcolm, Masson, and You." *Columbia Journalism Review* (March/April 1991): 38–44. (Discusses the ethical implications involved in psychoanalyst Jeffrey Masson's ten-million-dollar libel suit against journalist Janet Malcolm for allegedly deliberately misquoting him in a *New Yorker* article.)

Kroll, Barry. "How College Freshmen View Plagiarism." *Written Communication* 3(2): 2003–2021 (April 1988). (Summarizes the results of an Indiana University student survey on reasons for, and attitudes about, plagiarism. The survey revealed that students considered plagiarism to be a serious issue and tended to interpret it in terms of fairness to authors and other students, responsibility of students to do their own work, and respect for the ownership rights of others.)

———. "Why Is Plagiarism Wrong?" Paper presented to faculty and students at DePauw University (November 11, 1987). (Discusses different punitive strategies instructors use to deter plagiarism—moralistic prohibitions, prevention, and deterrence—based on assumptions that students knowingly engage in wrongdoing. Kroll suggests an alternative assumption—that most students plagiarize out of ignorance—and recommends a strategy of increasing education about what constitutes plagiarism.)

Peterson, M. J. "Everything You Ever Wanted to Know about Documentation (and Plagiarism), But Were Afraid to Ask, with Illustrations and Examples." Mimeographed paper (February 1983). (Gives examples of correct and incorrect uses of quotations and other written materials.)

Professional Codes of Ethics

Callahan, Daniel. "Should There Be an Academic Code of Ethics?" *Journal of Higher Education* 53(3): 335–341 (1982). (Argues against a code of ethics for academics based on the author's claim that codes are generally developed as responses to a state of disorder within a profession, are generally ineffective, and would be either too broad or too narrow to be useful.)

Fry, Amelia. "Reflections on Ethics." In *Oral History: An Interdisciplinary Anthology*, edited by David Dunaway and Willa Baum, pp. 150–161. Nashville, TN: American Association for State and Local History, 1984. (Discusses the collaborative nature of the oral history process, limitations of the ethical guidelines of the Oral History Association, researchers' obligations to narrators, and other ethical issues that arise in the process of oral historical research.)

Hoff-Wilson, Joan. "Access to Restricted Collections: The Responsibility of Professional Historical Organizations." *American Archivist* 46(4): 441–447 (Fall 1983). (Argues that the Organization of American Historians should reconsider its position on codes of ethics. Mentions problems to avoid in adopting codes of ethics and discusses recent trends in historical research methodology.)

Newton, Lisa H. "Lawgiving for Professional Life: Reflections on the Place of the Professional Code." In *Professional Ideals*, edited by Albert Flores, pp. 47–55. Belmont, CA: Wadsworth, 1988. (Argues that having a professional ethic is a prerequisite to being a profession and that codes of ethics preserve autonomy and rights for professionals.)

Schur, G. M. "Toward a Code of Ethics for Academics." *Journal of Higher Education* 53(3): 318–334 (1982). (Argues that academics need a code of ethics to ensure their accountability and outlines the elements of such a code.)

LONGER REFERENCES AND REFERENCES OF SECONDARY INTEREST

Plagiarism

Hurtado, Albert. "Historians and Their Employers: A Perspective on Professional Ethics." *Public Historian* 3(1): 47–51 (Winter 1986). (Argues in favor of a code of ethics for public historians.)

Mallon, Thomas. *Stolen Words: Forays into the Anguished Ravages of Plagiarism*. New York: Ticknor & Fields, 1989. (Discusses several recent cases of "high-visibility" plagiarism in literature and film.)

Shaw, Peter. "Plagiary." *American Scholar* 51: 325–327 (Summer 1981). (Discusses plagiarism in literature.)

Shea, John. "When Borrowing Becomes Burglary." *Currents* 13(1): 38–42 (January 1987). (Discusses plagiarism in college alumni magazines.)

Professional Codes

Abbott, Andrew. "Professional Ethics." *American Journal of Sociology* 88(5): 855–885 (March 1983). (An advanced analysis of five basic properties common to codes of professional ethics. Useful for graduate students exploring this issue in depth.)

American Historical Association. *Statement on Standards of Professional Conduct*. Washington, DC: American Historical Association, 1993. (This booklet contains the AHA's professional statements. In addition to the statement from which it derives its title, the booklet contains all of the guidelines on professional conduct and practice developed by the AHA's Professional Division. Included are the association's procedures for enforcing its standards of professional conduct and the following documents: "Statement on Plagiarism and Related Misuses of the Work of Other Authors," "Advisory Opinion Regarding the Harassment of Job Candidates," "Statement on Interviewing for Historical Documentation," "Statement on Discrimination and Harassment in Academia," "Advisory Opinion Regarding Conflict of Interest," and "Statement on Diversity in History Teaching.")

Orzack, Louis, and Simcoes, Annell, eds. *The Professions and Ethics: Views and Realities in New Jersey*. New Brunswick, NJ: Rutgers University Press, 1982. (Monograph summarizing Rutgers University's Professions Forum. Of particular relevance are contributions by Daniel Callahan ["Do Special Ethical Norms Apply to Professions?"] and Frank Fischer ["Ethical Codes: Sorting Out Rhetoric and Reality"].)

Williams, James. "Standards of Professional Conduct in California." *Public Historian* 8(1): 57–59 (Winter 1986). (Argues for a separate set of standards of professional conduct for public historians.)

2. *Chapter 18, Oral Historians Meet the Media*

SHORTER REFERENCES OF PRIMARY INTEREST

Allen, Barbara. "Re-creating the Past: The Narrator's Perspective in Oral History." *Oral History Review* 12: 1–12 (1984). (Discusses the differences between narrators' and researchers' views of history and the resulting damage to the integrity of history when it is divided into "usable" and "unusable" fragments.)

American Historical Association. "Statement on Interviewing for Historical Documentation." In *Statement on Standards of Professional Conduct*, pp. 17–19.

Borland, Katherine. " 'That's Not What I Said': Interpretive Conflict in Oral Narrative Research." In *Women's Words: The Feminist Practice of Oral History*, edited by Sherna Gluck and Daphne Patai, pp. 63–76. New York: Routledge, 1991. (Relates the author's "negotiations" with her narrator regarding an acceptable interpretation of the narrator's story [e.g., whether growing up as an independent woman in the nineteenth century meant the narrator was a "feminist"], and indicates the importance of researchers' sharing of their texts with narrators to avoid pitfalls that result from assuming shared meanings.)

Fry, Amelia. "Reflections on Ethics." In *Oral History: An Interdisciplinary Anthology*, edited by David Dunaway and Willa Baum, pp. 150–161. Nashville, TN: American Association for State and Local History, 1984. (Discusses the collaborative nature of the oral history process, limitations of the ethical guidelines of the Oral History Association, researchers' obligations to narrators, and a number of other ethical issues that arise in the process of doing oral historical research.)

Gluck, Sherna. "What's So Special about Women? Women's Oral History." In *Oral History: An Interdisciplinary Anthology*, pp. 221–237. (This landmark survey argues that women's oral history is inevitably a feminist endeavor. Discusses strategies for conducting oral histories, especially with older women.)

Hoffman, Alice. "Reliability and Validity in Oral History." In *Oral History: An Interdisciplinary Anthology*, pp. 67–73. (Discusses constructive ways to deal with the problems of reliability and validity that arise in the conduct of oral historical research, including suggestions for reconciling inconsistencies between testimony and documentary sources.)

Horowitz, Steven. "Review Essay: Crimes in Search of Evidence: Journalists' Use of Oral History." *Oral History Review* 17: 125–129 (Fall 1989). (Review essay of three political histories [*"Them": Stalin's Political Puppets, Endgame: The Fall of Marcos*, and *Enter the Dragon: China's Undeclared War against the U.S. in Korea 1950–51*]. Discusses the ways in which oral testimony may be manipulated to express the authors' biases.)

Lang, William, and Mercier, Laurie. "Getting It Down Right: Oral History's Reliability in Local History Research." *Oral History Review* 12: 81–99 (1984). (Proposes criteria for assessing the reliability of oral history, including consideration of the particular circumstances under which the material was gathered, the kinds of questions asked, and the rapport between narrator and interviewer. Also discusses particular problems in conducting oral historical research.)

May, William. "Doing Ethics: The Bearing of Ethical Theories on Fieldwork." *Social Problems* 27: 358–370 (1980). (Discusses the relevance of teleological, utilitarian, deontological, critical philosophical, and conventional ethical theories to fieldwork.)

Olsen, Karen, and Swopes, Linda. "Crossing Boundaries, Building Bridges: Doing Oral History among Working-Class Women and Men." In *Women's Words: The Feminist Practice of Oral History*, edited by Sherna Gluck and Daphne Patai, pp. 175–188. New York: Routledge, 1991. (On the basis of their oral history interviews with working-class women, the authors discuss the need to incorporate into oral historical research the significance of power relations and their links with gender, class, race, and ethnicity; to take account of the inherently unequal relationship between researcher and subject; and to take responsibility for making the relationship between the interview context and "real life" a meaningful one.)

Oral History Association. *Oral History Evaluation Guidelines*. Racine, WI: Wingspread Evaluation Conference, 1980. (Contains the Oral History Association's guidelines for interviewers and interviewees. Covers program and project design, ethical and legal aspects of interviewing, tape and transcript processing, and interview content and conduct.)

Patai, Daphne. "U.S. Academics and Third World Women: Is Ethical Research Possible?" *Women's Studies in Indiana* 15(1): 104–105 (November/December 1989). (Explores, from a feminist perspective, the ethical problems associated with affluent Western researchers conducting fieldwork involving Third World women. Discusses the importance of placing ethical principles within the context of the material realities of people's lives, and considers whether feminist researchers can avoid exploiting their subjects.)

———. "Ethical Problems of Personal Narratives, or, Who Should Eat the Last Piece of Cake?" *International Journal of Oral History* 8(1): 5–27 (1987). (Describes the social and economic power inequalities of subjects and oral historians and the need to develop methods to ameliorate the potential for exploitation.)

Thompson, Paul. "History and the Community." In *Oral History: An Interdisciplinary Anthology*, pp. 37–50. (One of Britain's first oral historians argues that oral history, particularly research on the history of the family, can serve to create positive social change and transform the social meaning of history by opening up traditional history to include different social groups, areas of inquiry, groups of researchers, and audiences.)

Thorne, Barrie. "You Still Takin' Notes? Fieldwork and Problems of Informed Consent." *Social Problems* 24: 284–297 (1980). (Considers problems faced by participant observers in complying with federal laws that require researchers to obtain informed consent from subjects when conducting research involving humans. Criticizes the assumptions underlying the principles of informed consent and argues that discussions of ethics and fieldwork should involve a critical look at informed consent.)

LONGER REFERENCES AND REFERENCES OF SECONDARY INTEREST

Bulmer, Martin, ed. *Social Research Ethics: An Examination of the Merits of Covert Participant Observation*. New York: Holmes & Meier, 1982. (Introduces the ethical problem of covert participant observation. Describes five case studies, contains several essays discussing ethical theory as it relates to covert methods, and describes the pros and cons of covert participant observation.)

Cataline, Chesley. "Ethics and the Human Dimension in Fieldwork." *Turkish Studies Association Bulletin* 12: 79–90 (1988). (Recounts several ethical dilemmas encountered by a fieldworker researching an indigent population in Turkey, including conflicts in values with subjects, getting access to information not freely offered, and keeping versus disclosing subjects' confidences about subjects such as wife and child abuse.)

Cutler, William, III. "Accuracy in Oral History Interviewing." In *Oral History: An Interdisciplinary Anthology*, pp. 79–86. (This early, foundational critique of the value of oral history research considers problems such as subjects' selective retention of information, self-deception, and forgetfulness. The article inspired a number of responses in defense of oral history.)

Dunaway, David, and Baum, Willa, eds. *Oral History: An Interdisciplinary Anthology*. Nashville, TN: American Association for State and Local History, 1984. (Collection of articles covering a wide variety of topics relating to oral history.)

Frisch, Michael, ed. *A Shared Authority: Essays on the Craft and Meaning of Oral and Public History*. Albany: State University of New York Press, 1990. (Collection of essays on various aspects of oral history. See especially chapter 4, "Oral History and the Presentation of Class Consciousness: the *New York Times* v. The Buffalo Unemployed," a version of which is excerpted in chapter 18 of this volume, and chapter 5, "Preparing Interview Transcripts for Documentary Publication: A Line-by-Line Illustration of the Editing Process," which considers the ethical issues involved in selectively incorporating oral interviews into documentary works.)

Gluck, Sherna, and Patai, Daphne, eds. *Women's Words: The Feminist Practice of Oral History*. New York: Routledge, 1991. (Collection of articles by feminist oral historians about the problems of conducting oral historical research.)

Langness, L. L., and Frank, Geyla. *Lives: An Anthropological Approach to Biography*. Novato, CA: Chandler & Sharp, 1981. (Chapter V, "Ethical and Moral Concerns," presents ethical problems in anthropological interviewing that are shared with oral history: balancing truthful reporting with the need to protect people's privacy, defining privacy, and the researcher's responsibilities in life-history studies.)

Patai, Daphne. *Brazilian Women Speak: Contemporary Life Stories*. New Brunswick, NJ: Rutgers University Press, 1988. (Collection of the oral histories of twenty "invisible" contemporary Brazilian women gathered by the author, who applies a feminist perspective to consider ethical issues she encountered in her fieldwork.)

Reece, Robert, and Siegel, Harvey. *Studying People: A Primer in the Ethics of Social

Research. Macon, GA.: Mercer University Press, 1986. (Covers a wide variety of ethical issues confronted when conducting research involving human subjects.)

Ryant, Carl. "The Public Historian and Business History: A Question of Ethics." *Public Historian* 8(1): 31–36 (Winter 1986). (Discusses the ethical problems involved in conducting public history research for business enterprises, including those common to oral historians generally and those particular to business historians.)

3. *Chapter 19, Limiting Access to Research Materials: The Case of the Dead Sea Scrolls*

SHORTER REFERENCES OF PRIMARY INTEREST

Coughlin, Ellen. "Opening of Dead Sea Scrolls Archive Underlines Problems That Can Complicate Access to Research Materials." *Chronicle of Higher Education* (October 2, 1991), pp. A6–A11. (Outlines problems created by lack of guidelines for access to scholarly research materials.)

Eisenman, Robert H., and Wise, Michael, eds. *The Dead Sea Scrolls Uncovered: The First Complete Translation and Interpretation of 50 Key Documents Withheld for Over 35 Years*. London and New York: Penguin Books, 1991. (The introduction to this interpretive volume sets forth the difficulties Eisenman faced when attempting to obtain access to unpublished scrolls and argues against exclusive access.)

Eisenman, Robert H., and Robinson, James M., eds. *A Facsimile Edition of the Dead Sea Scrolls*. 2 vols. Washington, DC: Biblical Archaeology Society, 1991. (Reproductions of photographs of previously unpublished texts. The introduction and publisher's foreword provide insights into the controversy surrounding exclusive access to the Dead Sea Scrolls. The photographs themselves demonstrate visually the difficulty involved in reconstructing scroll fragments.)

Levering, Eugene. "In Dead Sea Scrolls Session, SBL Issues a Statement on Access to Ancient Written Materials." *Religious Studies News* 7(1): 3–5 (January 1992). (Discusses the background to the Society of Biblical Literature's decision to issue its statement, "Access to Ancient Materials.")

Maranz, Felice. "Special Report: Unveiling the Treasures of Qumran, A New Generation for the Scrolls." *Jerusalem Report* II(3): 6–7 (November 7, 1991). (A one-page summary of the scrolls controversy and the decision to open access completely).

Nova. "Secrets of the Dead Sea Scrolls." Princeton, NJ: Films for the Humanities and Sciences, Inc., 1991. (Videotape.)

Shanks, Hershel. "Why the Dead Sea Scrolls Had to Be Released." *Chronicle of Higher Education* (October 2, 1991), pp. B1–B2. (A summary of the problems created by extensive delays in making newly discovered manuscripts and inscriptions available to the public. Also includes proposals for increasing scholars' access to ancient texts and for a code of conduct governing translations of texts.)

——, ed. "Dead Sea Scrolls Update: The Dead Sea Scroll Monopoly Must Be Broken." *Biblical Archaeology Review* 16: 44–46 (July/August 1990). (Describes unsuccessful efforts of biblical scholars to get access to Dead Sea Scrolls controlled by the official editing team and the team's inadequate rationale for denying access.)

Sheler, Jeffrey. "Can Ideas Be Held Hostage?" *U.S. News and World Report* (June 25, 1990), pp. 56–57. (Brief overview of the scrolls controversy, including a description of disputes in biblical interpretation that might be resolved by publication of the scrolls.)

LONGER REFERENCES AND REFERENCES OF SECONDARY INTEREST

Rogers, Jeffrey. "Needed: A Board to Handle Ancient Documents." *Chronicle of Higher Education* (October 30, 1991), p. 82. (Proposes that an international board of trustees be appointed to handle administration and assignment of newly discovered ancient documents.)

Shanks, Hershel. "Who Controls the Scrolls?" *Biblical Archaeology Review* 17(2): 52–53 (March/April 1991). (Provides a summary description of the major players on the scrolls team and their positions regarding access.)

Vermes, Geza. *The Dead Sea Scrolls in English*, 3d ed. London: Pelican, 1987. (Contains the text of certain nonbiblical texts from the Dead Sea Scrolls, with commentary.)

Case Index

This index provides a listing of the topics covered in each case and the relevant page numbers for each citation. The subjects listed directly following the chapter title are covered in each case or each section of the chapter. Subjects given after a case or subpart number are addressed in addition to the topics listed following the chapter title.

Chapter 3 (Biology: The Professional Scientist), 29–40
Case 1, 31–32: Acknowledgment; Authorship; Correcting Errors; Publishing; Replication/replicability of results
Case 2, 32–33: Correcting errors; Publishing
Part A: Replication/replicability of results
Part B: Clinical trials; Human subjects
Case 3, 33–34: Correcting errors; Fraud
Case 4, 34–35: Data/research material ownership; Data/research material sharing; Fraud; Intellectual property; Investigations; Misconduct; Omission of data; Replication/replicability of results; Sanctions for misconduct; Whistleblowing
Case 5, 36–37: Authorship; Data/research material sharing; Mentors; Priority (of right to research a subject)
Part C: Pressure to publish; Publishing
Part D: Data/research material ownership; Industry-university relations; Intellectual property
Part E: Industry-university relations
Case 6, 37–39: Confidentiality; Intellectual property; Priority (of right to research a subject)
Part A: Acknowledgment
Part B: Peer review
Case 7, 39: Error; Students (as subjects); Teaching
Case 8, 39–40: Independence of research; Editorial control over published research

Chapter 4 (Biology: Scientific Misconduct: What Is It and How Is It Investigated?), 41–50: Data alteration; Fraud; Intellectual property; Misconduct
Part A, 42–43
Part B, 43–45: Authorship; Fabrication; Intent to deceive; Omission of data
Part C, 45: Investigations
Part D, 45–46: Data/research material ownership; Data retention
Part E, 46–47: Fabrication; Pressure to publish; Reporting "expected" results
Part F, 47–48: Confidentiality; Investigations
Part G, 48: Investigations; Sanctions for misconduct
Part H, 49
Part I, 49–50: Intent to deceive; Investigations; Publishing
Part J, 50: Investigations; Sanctions for misconduct

Chapter 5 (Biology: Authorship and the Use of Scientific Data), 51–55
Case 1, 52: Authorship; Citation; Data alteration; Plagiarism

Case 2, 52: Authorship; Citation; Permission (to use another's work); Plagiarism
Case 3, 52: Authorship; Citation; Permission (to use another's work); Plagiarism
Case 4, 53: Authorship; Citation; Permission; Plagiarism
Case 5, 53: Citation; Data alteration; Permission; Teaching
Case 6, 54: Permission; Pressure to publish [in Instructional Notes]; Plagiarism
Case 7, 54–55: Citation; Data/research material ownership; Intellectual property

Chapter 6 (Biology: Data Alteration in Scientific Research), 56–61
Case 1, 56: Data alteration; Teaching
Case 2, 57: Data alteration; Omission of data; Teaching
Case 3, 57–58: Data alteration; Omission of data
Case 4, 58–59: Data alteration
Case 5, 59: Omission of data
Case 6, 59: Omission of data
Case 7, 60: Clinical trials [in Instructional Notes]; Omission of data
Case 8, 60: Authorship; Data alteration; Graduate students; Interpreting data

Chapter 7 (Biology: The Ethics of Genetic Screening and Testing), 62–71: Genetic screening and testing; Responsibility for the effects of research
Case 1, 63–64: Beta-thalassemia; Confidentiality; Error; Genetic counseling; Informed consent; Privacy
Case 2, 65: Confidentiality; Privacy; Wrongful birth; Wrongful life [in Instructional Notes]
Case 3, 65–67: Confidentiality; Employment screening; Hypersusceptibility; Privacy
Case 4, 67–68: Genetic counseling; Huntington disease; Minors

Chapter 8 (Biology: Ethics and Eugenics), 72–84: Eugenics; Responsibility for the effects of research; Pedigrees
Case 1, 74–76: Embryo selection [in Instructional Notes]; Military heroism
Case 2, 76–79: Feeblemindedness; Intelligence tests; Jukes family; Kallikak family
Case 3, 79–81: Kallikak family; Feeblemindedness; Interviewing

Chapter 9 (Psychology: Ethical Issues in Animal Experimentation), 87–98: Animal research; Animal rights; Civil disobedience; Euthanization of animals; Institutional animal care and use committees
Case 1, 88–91
Case 2, 91–93

Chapter 10 (Psychology: Research Involving Human Subjects), 99–101: Deception (of human subjects); Human subjects; Informed consent

Chapter 11 (Psychology: Research Involving Human Subjects: The Administration of Alcohol), 105–111: Alcoholism; Deception (of human subjects); Human subjects; Informed consent; Recruiting human subjects; Risk of harm
Recruitment Scenario 1, 107–108
Recruitment Scenario 2, 108–109
Recruitment Scenario 3, 109
Recruitment Scenario 4, 109–110

Chapter 12 (Psychology: The Ethics of Deception in Research), 112–118: Debriefing (of human subjects); Deception (of human subjects); Human subjects; Informed consent; Risk of harm; Risk versus benefit
Case 1, 113–114
Case 2, 114
Case 3, 115–116

Case 4, 116–117
Case 5, 117–118

Chapter 13 (Psychology: Misconduct in Science), 119–124: Animal research; Investigations [in Instructional Notes]; Misconduct
Part A, 120–121: Conflicts of interest; Industry-university relations
Part B, 121: Institutional animal care and use committees [in Instructional Notes]
Part C, 121: Omission of data; Publishing
Part D, 122: Omission of data; Publishing
Part E, 122: Fabrication; Pressure to publish; Reporting "expected" results
Part F, 123: Omission of data; Publishing
Part G, 123–124: Authorship; Citation

Chapter 14 (Psychology: Science and Coercion), 125–136: Coercion; Human subjects; Informed consent; Institutional review boards
Case 1, 126–128: Debriefing (of human subjects); Students (as subjects); Teaching
Case 2, 128–130: Students (as subjects)
Case 3, 130–131: Clinical trials; Students (as subjects)
Case 4, 131–132: Students (as subjects)
Case 5, 132–134: Responsibility for the effects of research

Chapter 15 (Psychology: Behavior Control), 137–144: Behavior; Control; Informed consent; Responsibility for the effects of research
Case 1, 138–139: Alcoholism
Case 2, 139–140: Alcoholism; Clinical trials; Human subjects
Case 3, 140: Alcoholism; Coercion
Case 4, 141–142: Minors
Case 5, 142: Human subjects; Minors
Case 6, 142–143: Minors
Case 7, 143–144

Chapter 16 (History: The Historian's Code of Ethics), 147–155: Sanctions for misconduct
Case 1, 148–151: Codes of ethics; Misuse of another's writings; Plagiarism
Case 2, 151–153: Dissertations; Investigations; Plagiarism; Whistleblowing
Case 3, 153–154: Plagiarism; Undergraduate students
Case 4, 154–155: Dissertations; Graduate students; Misuse of another's writings

Chapter 17 (History: The Use and Interpretation of Historical Documents), 156–159: Citation; Documents; Plagiarism

Chapter 18 (History: Oral Historians Meet the Media), 160–172: Confidentiality; Deception (of human subjects) [in Instructional Notes]; Editorial control over published research; Interviewing; Oral history; Responsibility for the effects of research

Chapter 19 (History: Limiting Access to Scholarly Materials: the Case of the Dead Sea Scrolls), 173–181: Access to research materials (humanities); Dead Sea Scrolls; Intellectual property; Priority (of right to research a subject)
Part A, 173–175: "Authorized" publications
Part B, 175–176: Stealing research materials
Part C, 176–177: Data/research material ownership
Part D, 178–179: "Authorized" publications; Codes of ethics
Part E, 179–180

Chapter 20 (History: Faculty-Graduate Student Relations), 182–186: Faculty-graduate student relations; Graduate students
Case 1, 182–183: Acknowledgment; Coercion; Intellectual property; Stealing ideas
Case 2, 183–184: Research assistants
Case 3, 184: Coercion; Mentors
Case 4, 185: Acknowledgment; Intellectual property; Mentors; Stealing ideas
Case 5, 185–186: Acknowledgment; Research assistants

Chapter 21 (History: Intellectual Property), 187–192: Intellectual property; Misuse of another's writings
Case 1, 187–188: Dissertations; Footnote mining; Journal editors; Peer review
Case 2, 188–189: Citation; Dissertations; Footnote mining; Stealing ideas
Case 3, 189: Citation; Dissertations; Footnote mining; Stealing ideas
Case 4, 190: Access to research materials; Dissertations; Priority (of right to research a subject)
Case 5, 190–191: Citation; Dissertations; Footnote mining
Case 6, 191: Access to research materials; Documents; Priority (of right to research a subject)
Case 7, 191–192: Peer review

General Index

Entries for some subject headings include listings of cases or parts of cases that specifically address the topic.

Academic freedom, 177
Access to research data/materials. *See* Data/research material sharing (sciences)
Access to research materials (humanities)
—*case studies:*
history, 173–181 (ch. 19), 190, 191 (ch. 21, Cases 4 and 6)
Acknowledgment
—*case studies:*
biology, 31–32, 37 (ch. 3, Case 1 and Case 6 [part B]); history, 182–183, 185, 185–186 (ch. 20, Cases 1, 4, and 5)
Act deontology, 20–21
Act utilitarianism, 16–17
AHA. *See* American Historical Association
Alcohol, 99–103
Alcoholism, 87, 138
—*case studies:*
psychology, 105–111 (ch. 11), 138–140 (ch. 15, Cases 1–3)
American Historical Association, 14, 147
—*Statement on Interviewing for Historical Documentation*, 232–233
—*Statement on Plagiarism and Related Misuses of the Work of Other Authors*, 148–150
—*Statement on Standards of Professional Conduct*, 161, 175, 179–180
American Psychological Association, 14, 103, 125–126
—*Ethical Principles in the Conduct of Research with Human Participants*, 103
—*Guidelines for Research Involving Human Subjects*, 103, 218–219
Ancient Biblical Manuscript Center, 176–177
Animal care and use committees. *See* Institutional animal care and use committees
Animal care and use regulations, 94–96
Animal research, 14
—*case studies:*
psychology, 87–98 (ch. 9), 121–124 (ch. 13, parts B-G)
Animal rights
—*case studies:*
psychology, 87–98 (ch. 9)
APA. *See* American Psychological Association
"Authorized" publications
—*case studies:*
history, 173–175, 178 (ch. 19, parts A and D)
Authorship
—*case studies:*
biology, 31–32, 36–37 (ch. 3, Cases 1 and 5), 43–45 (ch. 4, part B), 52–53 (ch. 5, Cases 1–4), 60–61 (ch. 6, Case 8); psychology, 123–124 (ch. 13, part G)
Autonomy, 125. *See also* Respect for persons
Aversive stimuli, 137

Bechtel, Elizabeth Hay, 176–177
Behavior control
—*case studies:*
psychology, 137–144 (ch. 15)
Behavior modification. *See* Behavior control
Belmont Report, The, 99, 103
Beneficence, 14, 99
Bentham, Jeremy, 15
Beta-thalassemia
—*case studies:*
biology, 63–64 (ch. 7, Case 1)
Binet test, 77
Biotechnology. *See* Industry-university relations
Buck, Carrie, 72

Casuistry, 15, 21–23, 25, 51, 56
Character. *See* Virtue ethics

Citation
—*case studies:*
biology, 52–53, 54–55 (ch. 5, Cases 1–5 and 7); psychology, 123 (ch. 13, part G); history, 156–159 (ch. 17), 188–189, 189, 190–191 (ch. 21, Cases 2, 3 and 5)
Civil disobedience, 88
—*case studies:*
psychology, 87–98 (ch. 9)
Clinical trials, 223
—*case studies:*
biology, 33 (ch. 3, Case 2, part B), 203–204 (Instructional Notes for ch. 6, Case 7); psychology, 130–131 (ch. 14, Case 3), 139–140 (ch. 15, Case 2)
Codes of ethics, 14, 24
—*case studies:*
history, 148–151 (ch. 16, Case 1), 178 (ch. 19, part D)
Coercion
—*case studies:*
psychology, 125–136 (ch. 14); history, 182–183, 184 (ch. 20, Cases 1 and 3)
Colon cancer, 63
Confidentiality
—*case studies:*
biology, 37–39 (ch. 3, Case 6), 47–48 (ch. 4, part F), 63–71 (ch. 7)
Conflicts of interest, 14, 120
—*case studies:*
psychology, 120–121 (ch. 13, part A)
Consent. *See* Informed consent
Consequentialism, 4, 15–17, 18, 25, 95, 203. *See also* Utilitarianism
Contractarian ethical theories, 96
Cooley's anemia. *See* Beta-thalassemia
Correcting errors, 30–31
—*case studies:*
biology, 31–34 (ch. 3, Cases 1–3), 60 (ch. 6, Case 7)
Credit (for work or ideas). *See* Acknowledgment
Cystic fibrosis, 63, 79

Data alteration, 120
—*case studies:*
biology, 41–50 (ch. 4), 52, 53 (ch. 5, Cases 1 and 5), 56–59, 60–61 (ch. 6, Cases 1–4 and 8)
Data/research material ownership
—*case studies:*
biology, 37 (ch. 3, Case 5, part D), 45–46 (ch. 4, part D), 54–55 (ch. 5, Case 7); history, 173–181 (ch. 19)
Data/research material sharing (sciences). *See also* Access to research materials (humanities)
—*case studies:*
biology, 34–36, 36–37 (ch. 3, Cases 4 and 5), 45–46 (ch. 4, part D); history, 176–177 (ch. 19, part C)
Data retention
—*case studies:*
biology, 45–46 (ch. 4, part D)
Dead Sea Scrolls
—*case studies:*
history, 173–181 (ch. 19)
Debriefing (of human subjects), 102
Deception (of human subjects)
—*case studies:*
psychology, 99–104 (ch. 10), 105–111 (ch. 11), 112–118 (ch. 12); history, 230–232 (Instructional Notes for ch. 18)
Deontology, 4, 15, 16, 17–21, 25, 95–96
Dissertations, 187
—*case studies:*
history, 151–153, 154–155 (ch. 16, Cases 2 and 4), 187–191 (ch. 21, Cases 1–5)
Disulfiram, 137–140
Documents
—*case studies:*
history, 156–159 (ch. 17), 191 (ch. 21, Case 6)
Down syndrome, 62, 73
Duties, 18, 19–20

Editorial control over published research
—*case studies:*
biology, 39–40 (ch. 3, Case 8); psychology, 120–121 (ch. 13, part A); history, 160–172 (ch. 18)
Eisenmann, Robert, 234
Embryo selection
—*case studies:*
biology, 212 (Instructional Notes for ch. 8, Case 1)
Employment screening
—*case studies:*
biology, 65–67 (ch. 7, Case 3)
Error
—*case studies*
biology, 39 (ch. 3, Case 7), 63–64 (ch. 7, Case 1)
Ethical theory, 4–5, 13–26, 14–15

Eugenics
—*case studies:*
biology, 72–84 (ch. 8)
Euthanization of animals
—*case studies:*
psychology, 87–98 (ch. 9, Cases 1 and 2)

Fabrication, 120. *See also* Reporting "expected" results
—*case studies:*
biology, 43–45, 46–47 (ch. 4, parts B and E); psychology, 122 (ch. 13, part E)
Faculty-graduate student relations. *See also* Graduate students
—*case studies:*
history, 182–186 (ch. 20)
Falsification. *See* Data alteration
Feeblemindedness
—*case studies:*
biology, 77–81 (ch. 8, Cases 2 and 3)
Footnote mining
—*case studies:*
history, 187–189, 190–191 (ch. 21, Cases 1–3 and 5)
Fraud, 14, 29–30, 31
—*case studies:*
biology, 33–35 (ch. 3, Cases 3 and 4), 41–50 (ch. 4)
Frisch, Michael, 159

Genetic counseling
—*case studies:*
biology, 63–64, 67–68 (ch. 7, Cases 1 and 4)
Genetic screening and testing
—*case studies:*
biology, 62–71 (ch. 7)
Genovese, Kitty, 222
Goddard, Henry H., 80, 81; summary of Kallikak family study, 81–83, 214
Government regulations, 5, 14, 125, 131, 134
Graduate students, 29. *See also* Faculty-graduate student relations
—*case studies:*
biology, 60–61 (ch. 6, Case 8); history, 154–155 (ch. 16, Case 4), 182–186 (ch. 20)

HD. *See* Huntington disease
Hedonic utilitarianism, 16
Holmes, Oliver Wendell, 72
Human subjects, 14, 230. *See also* Recruiting human subjects
—*case studies:*
biology, 33 (ch. 3, Case 2, part B); psychology, 99–104 (ch. 10), 105–111 (ch. 11), 112–118 (ch. 12), 125–136 (ch. 14), 139–140, 142 (ch. 15, Cases 2 and 5)
Huntington disease, 63, 79, 132, 205
—*case studies:*
biology, 67–68 (ch. 7, Case 4)
Huntington Library, 176–177, 179
Hypersusceptibility
—*case studies:*
biology, 65–67 (ch. 7, Case 3)

IACUC. *See* Institutional animal care and use committees
Inclusion of data (in publications). *See* Omission of data
Independence of research, 31
—*case studies:*
biology, 39–40 (ch. 3, Case 8); psychology, 120–121 (ch. 13, part A)
Industry-university relations, 14
—*case studies:*
biology, 37 (ch. 3, Case 5, parts D and E); psychology, 120–121 (ch. 13, part A)
Informed consent, 125
—*case studies:*
biology, 63–64 (ch. 7, Case 1); psychology, 99–104 (ch. 10), 105–111 (ch. 11), 112–118 (ch. 12), 125–136 (ch. 14), 137–144 (ch. 15); history, 230, 231, 233 (Instructional Notes for ch. 18)
Institutional animal care and use committees, 87, 121
—*case studies:*
psychology, 87–98 (ch. 9, Cases 1 and 2), 121, 222–223 (ch. 13, part B)
Institutional review boards, 99, 112–113, 125–126
—*case studies:*
psychology, 99–104 (ch. 10), 105–111 (ch. 11), 112–118 (ch. 12), 125–136 (ch. 14), 139–140 (ch. 15, Case 2)
Intellectual property, 14, 31
—*case studies:*
biology, 34–35, 36, 37–39 (ch. 3, Case 4, Case 5 [part B] and Case 6), 41–50 (ch. 4), 54–55 (ch. 5, Case 7); history, 173–181 (ch. 19), 182–183, 185 (ch. 20, Cases 1 and 4), 187–192 (ch. 21)
Intelligence tests
—*case studies:*
biology, 77–79 (ch. 8, Case 2)
Intent to deceive, 5

Intent to deceive—(*continued*)
—*case studies:*
biology, 43–45, 49–50 (ch. 4, parts B and I)
Interpreting data
—*case studies:*
biology, 60–61 (ch. 6, Case 8)
Interviewing
—*case studies:*
biology, 79–83 (ch. 8, Case 3); history, 160–172 (ch. 18)
Investigations, 31
—*case studies:*
biology, 34–35 (ch. 3, Case 4), 45, 47–48, 48, 49–50, 50 (ch. 4, parts C, F, G, I, and J); psychology, 121, 222–223 (ch. 13, part B and Instructional Notes); history, 151–153 (ch. 16, Case 2)
IRBs. *See* Institutional review boards
Israeli Antiquities Authority, 174, 175, 176–177, 179

Job screening. *See* Employment screening
Journal editors, 187
—*case studies:*
history, 187–188 (ch. 21, Case 1)
Jukes family
—*case studies:*
biology, 77–79 (ch. 8, Case 2)
Justice, 14, 99

Kallikak family, 77, 79–83; study summary, 81–83
—*case studies:*
biology, 79–83 (ch. 8, Case 3)
Kant, Immanuel, 15
Kinship, 96

Mental retardation. *See* Feeblemindedness
Mentors, 5–6, 31, 151
—*case studies:*
biology, 36–37 (ch. 3, Case 5); history, 184, 185 (ch. 20, Cases 3 and 4)
Milgram experiments, 218
Military heroism
—*case studies:*
biology, 74–76 (ch. 8, Case 1)
Mill, James, 15
Mill, John Stuart, 15
Minors
—*case studies:*
biology, 67–68 (ch. 7, Case 4); psychology, 141–143 (ch. 15, Cases 4–6)
Misconduct, 14
—*case studies:*
biology, 34–35 (ch. 3, Case 4), 41–50 (ch. 4); psychology, 119–124 (ch. 13)
Misconduct policies, 5, 9, 14, 193, 200, 205, 222–223
Misuse of another's writings
—*case studies:*
history, 148–151, 154–155 (ch. 16, Cases 1 and 4), 187–189, 190–191 (ch. 21, Cases 1–3 and 5)
Moral principles, 14, 16, 18, 20–21, 96; use in casuistry, 21, 22
Moral rules, 18, 20–21; use in virtue ethics, 23
Muscular dystrophy, 63

National Advisory Council on Alcohol Abuse and Alcoholism, 219; *Recommended Council Guidelines*, 219, 221–222
New York Times, 160–172

Omission of data (in reports/publications), 5
—*case studies:*
biology, 34–35 (ch. 3, Case 4), 43–45 (ch. 4, part B), 57, 57–58, 59–60 (ch. 6, Cases 2, 3, 5–7)
Oral history
—*case studies:*
history, 160–172 (ch. 18)
Organic unity theory, 96
Outliers. *See* Omission of data

Pedigrees
—*case studies:*
biology, 72–84 (ch. 8)
Peer review
—*case studies:*
biology, 38–39 (ch. 3, Case 6, part B); history, 187–188, 191–192 (ch. 21, Cases 1 and 7)
Permission (to use another's work)
—*case studies:*
biology, 52–55 (ch. 5, Cases 2–7)
Phenylketonuria, 62–63, 73
PKU. *See* Phenylketonuria
Plagiarism, 5, 14, 147. *See also* American Historical Association: *Statement on Plagiarism and Related Misuses of the Work of Other Authors*; Stealing ideas
—*case studies:*
biology, 52, 53, 54 (ch. 5, Cases 1–4 and 6); history, 148–154 (ch. 16, Cases 1–3), 156–159 (ch. 17)

Pluralist utilitarianism, 16
Policies. *See* Misconduct policies
Positive reinforcement, 137
Preference utilitarianism, 16
Pressure to publish, 197–198
—*case studies:*
biology, 36–37 (ch. 3, Case 5, part C), 46–47 (ch. 4, part E), 197–198 (Instructional Notes for ch. 5, Case 6); psychology, 122 (ch. 13, part E)
Principle of utility, 16
Priority (of right to research a subject)
—*case studies:*
biology, 36–39 (ch. 3, Cases 5 and 6); history, 173–181 (ch. 19), 190, 191 (ch. 21, Cases 4 and 6)
Privacy, 103
—*case studies:*
biology, 63–67 (ch. 7, Cases 1–3); history, 230–233 (Instructional Notes for ch. 18)
Publishing
—*case studies:*
biology, 31–32, 32–33, 36–37 (ch. 3, Cases 1, 2, and 5), 49–50 (ch. 4, part I); psychology, 121, 122, 123 (ch. 13, parts C, D, and F)

Randomized clinical trials. *See* Clinical trials
Recruiting human subjects, 125
—*case studies:*
psychology, 105–111 (ch. 11)
Regulations. *See* Government regulations
Replication/replicability of results
—*case studies:*
biology, 31–32, 32–33, 34–35 (ch. 3, Cases 1, 2 [part A], and 4)
Reporting data. *See* Omission of data
Reporting "expected" results. *See also* Fabrication
—*case studies:*
biology, 46–47 (ch. 4, part E); psychology, 122 (ch. 13, part E)
Research assistants, 182
—*case studies:*
history, 182–183, 183–184, 185–186 (ch. 20, Cases 1, 2, and 5)
Research misconduct. *See* Misconduct
Respect for persons, 14, 99, 112. *See also* Autonomy
Responsibility for the effects of research, 14, 225–226
—*case studies:*
biology, 62–71 (ch. 7), 72–86 (ch. 8); psychology, 132–134 (ch. 14, Case 5); history, 160–172 (ch. 18)
Rest, James, 10
Retinoblastoma, 38, 63, 78, 208
Reviewing manuscripts for publication. *See* Peer review
Rights, 19–20
Risk of harm, 105–106, 112–113
—*case studies:*
psychology, 105–111 (ch. 11), 112–118 (ch. 12)
Risk versus benefit, 99, 103
—*case studies:*
psychology, 112–118 (ch. 12)
Rule utilitarianism, 16, 17

Sanctions for misconduct, 147, 198, 202
—*case studies:*
biology, 34–35 (ch. 3, Case 4, part A), 48, 50 (ch. 4, parts G and J); history, 147–154 (ch. 16, Cases 1–3)
Scientific importance, 37, 90, 198, 199
Scientific misconduct. *See* Misconduct
Sharing data. *See* Data/research material sharing
Situation ethics, 15, 20–21. *See also* Act deontology
Society of Biblical Literature, 178
Speciesism, 87
Statistical significance, 31
Stealing ideas. *See also* Plagiarism
—*case studies:*
history, 182–183, 185 (ch. 20, Cases 1 and 4), 188–189 (ch. 21, Cases 2 and 3)
Stealing research materials
—*case studies:*
history, 175–176 (ch. 19, part B)
Students (as subjects)
—*case studies:*
biology, 39 (ch. 3, Case 7); psychology, 126–132 (ch. 14, Cases 1–4)

Teaching, 3–12, 197, 200–201
—*case studies:*
biology, 39 (ch. 3, Case 7), 53 (ch. 5, Case 5), 56, 57 (ch. 6, Cases 1 and 2); psychology, 126–128 (ch. 14, Case 1)
Theft of research materials. *See* Stealing research materials
Theory. *See* Ethical theory
Training School for Backward and Feeble-minded Children (Vineland, NJ), 81

Undergraduate students, 194, 199, 226
—*case studies:*
history, 153–154 (ch. 16, Case 3)
Undue influence. *See* Recruiting human subjects
United States Department of Agriculture: animal care and use regulations, 94–96
United States Department of Health and Human Services: human subjects regulations, 134–135; human subjects regulations for research involving children, 224–225
United States President's Commission for the Study of Ethical Problems in Medicine and Biomedical and Behavioral Research: *Screening and Counseling for Genetic Conditions*, 68–71
Utilitarianism, 4, 15–17, 95. *See also* Consequentialism

Virtue ethics, 15, 23–25, 25

Watts, Dorothy, 159, 170–171
Welfare utilitarianism, 16
Whistleblowing, 7–9
—*case studies:*
biology, 34–35 (ch. 3, Case 4); history, 151–153 (ch. 16, Case 2)
Wilberforce, William, 156–159
Witnessing misconduct. *See* Whistleblowing
Wrongful birth, 209
—*case studies:*
biology, 65 (ch. 7, Case 2)
Wrongful life, 209
—*case studies:*
biology, 209 (Instructional Notes for ch. 7, Case 2)

Robin Levin Penslar is Special Assistant to the Vice President for Research at Indiana University, following many years as a Research Associate for the Poynter Center for the Study of Ethics and American Institutions. She is the principal author and editor of *Protecting Human Research Subjects: Institutional Review Board Guidebook.*